Troubleshooting and Repairing Major Appliances

To my daughter, Brandi.

Troubleshooting and Repairing Major Appliances

Eric Kleinert

McGraw-Hill

New York San Francisco Washington, D.C. Auckland Bogotá
Caracas Lisbon London Madrid Mexico City Milan
Montreal New Delhi San Juan Singapore
Sydney Tokyo Toronto

McGraw-Hill

A Division of The McGraw-Hill Companies

©1995 by **McGraw-Hill, Inc**.
Published by TAB Books, a division of McGraw-Hill, Inc.

pbk   3 4 5 6 7 8 9 DOC/DOC 9 0 0 9 8
hc   4 5 6 7 8 9 DOC/DOC 9 0 0 9 8

Library of Congress Cataloging-in-Publication Data
Kleinert, Eric.
      Troubleshooting and repairing major appliances / by Eric Kleinert.
         p. cm.
      Includes index.
      ISBN 0-07-035079-5 (P)   ISBN 0-07-035078-7 (H)
      1. Household appliances, Electric—Maintenance and repair.
   I. Title.
   TA350.K654 1994
   683'.83'0288—dc20
                                                              94-34891
                                                              CIP

Acquisitions editor: April D. Nolan
Editorial team: Bill Taylor, N3LRA, Editor
                        Andrew Yoder, Managing Editor
                        Joanne Slike, Executive Editor
                        Joann Woy, Indexer
Production team: Katherine G. Brown, Director
                        Ollie Harmon, Coding
                        Rose McFarland, Desktop Operator
Design team: Jaclyn J. Boone, Designer
                        Katherine Lukaszewicz, Associate Designer

TAB1
0350795

Contents

PART TWO
APPLIANCE SELECTION, SERVICE, INSTALLATION, AND PREVENTIVE MAINTENANCE PROCEDURES

Acknowledgments

Major Appliance Consumer Action Panel (MACAP).

I thank April Nolan, and the staff of TAB/McGraw-Hill, for giving me the opportunity to write this book.

For his technical review of the manuscript, and for his corrections and suggestions, I thank the technical editor, William Taylor.

To my beloved, long-suffering wife Eileen, for all of the things that she has had to sacrifice, as I organized and wrote this book.

Lastly, I thank my daughter, Brandi, for her wisdom, experience, and assistance.

Introduction

Contrary to popular belief, appliance technicians don't perform magic when fixing broken machines. A disappearing act isn't required to master the operation and mechanics of a major appliance. All that is needed are the abilities to think clearly and use hand tools.

As a novice servicer, you might have to look up terms in the glossary, or consult the component descriptions and diagnostic charts. However, you'll be amply rewarded for this effort when a major appliance is fixed for the first time. This book is also informative and useful for the experienced servicer and repair technician.

This book was designed in two parts: general and specific. *Part One: the fundamentals of service*, covers the basic skills. The problems of selection, purchase, and installation of new major home appliances are introduced in the first chapter. Also included is information on manufacturer's warranties, use and care manuals, and where the consumers can get help if it is needed.

From chapters 2 through 6, a complete guide is laid out on your journey into the repair of major appliances. The complete coverage of safety precautions (for the installation, operation, and repair of major appliances); the correct tools needed for the job; and the basic approach both before, and during, the repair of major appliances; is just the beginning.

Everything you will ever need to know about electricity is provided in chapter 5. And, chapter 6 furnishes a description of the operation and location of the parts that are used in today's appliances.

Your journey continues through the land of major appliance repair in *Part Two: Appliance service, installation and preventive maintenance procedures.* Chapters 7 through 14 cover domestic dishwashers, garbage disposals, water heaters, automatic washers, automatic dryers, ranges/ovens, refrigerators, freezers, and automatic ice makers.

The step-by-step procedures for the testing and replacing of parts, troubleshooting charts, preventive maintenance procedures, and appliance installation procedures, are both thorough and comprehensive.

The troubleshooting information covers the various types of major home appliances in general terms, rather than as specific models, in order to present a broad overview of service techniques. The pictures and illustrations are used for demonstration purposes, and to clarify the description of how to service them. They do not reflect a particular manufacturer's reliability.

This book should provide you with a basic understanding of the operation of, and the problems common to, major appliances. This text will guide you as you complete your venture into the world of major appliances, and in becoming a qualified technician.

Part one

Fundamentals of service

1
Selecting, purchasing, and installing major home appliances

Both the technician and the consumer will appreciate this chapter because it addresses some of the problems of selection, acquisition, and installation of major home appliances. It is also pertinent that technicians familiarize themselves with the features and functions of the appliances, which are stated in every new appliance's use and care manual. This will be essential when diagnosing problems with the diverse types of appliances.

This information will provide proper planning and a better understanding of major home appliances, including appliance warranties, and where to get help when it is needed.

In today's market, major home appliances are manufactured to meet the needs of the average person. Remember that price should not be the most influential factor when choosing an appliance. Physical and mental limitations should also be considered when selecting the appliance that will be juxtaposed with the consumer's needs.

RANGES, COOKTOPS, AND OVENS

The domestic range was designed as a multi-purpose cooking appliance. It consists of a surface area, with heating elements on the top, to cook the food. The oven cavity is used for baking food at a set temperature. Within the same oven cavity, the broiling of food is also incorporated.

Domestic ranges are available in a choice of electric or gas, in sizes ranging from 20 inches to 40 inches in width, and with a wide selection of configurations and colors.

Freestanding ranges

The freestanding range stands between two base cabinets, or sometimes at the end of a cabinet line. These ranges are available in 20-, 24-, 30-, 36-, and 40-inch widths

(Fig. 1-1). Some 40-inch models have two ovens. Designs include gas burners, standard electric heating elements, and a glass cooktop with concealed electric elements underneath. The controls might be located on the rear console, or across the front. Oven cleaning systems include self-cleaning[1], continuous-cleaning[2], and standard-manual cleaning.

1-1
The freestanding range is available in 20-, 24-, 30-, 36-, and 40-inch widths and a variety of colors.

Listed below are some advantages and disadvantages of freestanding ranges (Fig. 1-1):

Advantages
- Freestanding ranges are generally less expensive than other types. Prices vary with features.
- Freestanding ranges can be moved when the family moves.
- Most models have some center space for placing utensils.
- Front controls can be reached easily from a seated position.
- Bottom drawer adds to kitchen storage space.

Disadvantages
- Rear console controls are virtually impossible for a seated person.
- Low broiler in gas ranges is less accessible, from a seated position, than the oven broiler in electric ranges.

Slide-in ranges and drop-in ranges

Slide-in ranges and drop-in ranges might not always be available in all sizes: 30 inch is the most common (Fig. 1-2). However, drop-in ranges are also available in a 27 inch width. Designs include gas burners, standard electric heating elements, and a glass cooktop with concealed electric elements underneath. The oven and cooktop controls are usually located across the front of the range. Some models have the

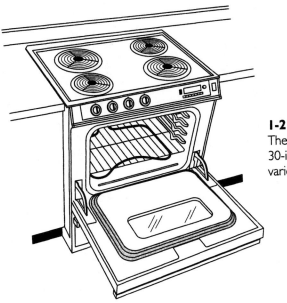

1-2
The slide-in range with a 30-inch width is available in a variety of colors.

cooktop controls along their side. They are an excellent choice for an island or peninsula shaped counter, because they are flush with the surrounding counter. These ranges can also be installed overlapping the adjacent countertop edges, eliminating dirt-catching gaps. A drop-in range either hangs from a countertop, or sits on a low cabinet base; and a slide-in range sits on the floor.

Listed below are some advantages and disadvantages of slide-in ranges (Fig. 1-2) and drop-in ranges (Fig. 1-3):

Advantages
- Can be installed a few inches higher or lower than a freestanding range.
- Controls can be reached by most cooks.

Disadvantages
- Built into kitchen and generally cannot be moved when the family relocates.
- Requires installation by a carpenter and an electrician.

Built-in cooktops

Built-in cooktops are set into a countertop and are made in various sizes, from 15 to 48 inches wide. Built-in cooktop designs include gas burners, standard electric heating elements, and a glass cooktop with concealed electric elements underneath (Figs. 1-4 and 1-5). They might have side or front controls, and might be of modular design. Special plug-in cooking accessories are also available. Listed below are some advantages and disadvantages of built-in cooktops:

Advantages
- Can be installed at the most convenient height for the cook.
- Side or front controls are easily reached by most cooks.
- Counter installation provides open space below the cooktop.

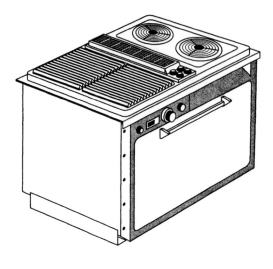

I-3
The drop-in range is most commonly available with a 30-inch width most common. It is available in a variety of colors.

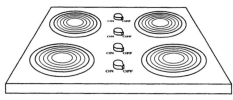

I-4
The electric built-in cooktops are available with one, two, or four heating elements and a variety of colors.

I-5
The electric built-in cooktop downdraft model. The unit has a fan in between the surface elements, which draws the cooking fumes into it and out of the kitchen.

Disadvantages
• Built into kitchen and generally cannot be moved when the family relocates.
• Requires installation by a carpenter and an electrician.

Built-in ovens

Built-in ovens usually have one oven cavity, but models with two oven cavities are also available. In the double oven cavity model, the second oven might be either a conventional, or a microwave, oven (Figs. 1-6 and 1-7). Built-in ovens are available in 24, 27, and 30 inch widths. They are designed to operate on either gas or electric. Their height varies, depending upon whether they are single or double oven units. Oven cleaning systems available include self-cleaning[1], continuous-cleaning[2] and standard manual-cleaning. Two-oven models also offer conventional ovens with two different cleaning systems. Built-in ovens, with the microwave feature, are available with either solid state microcomputer or electromechanical controls.

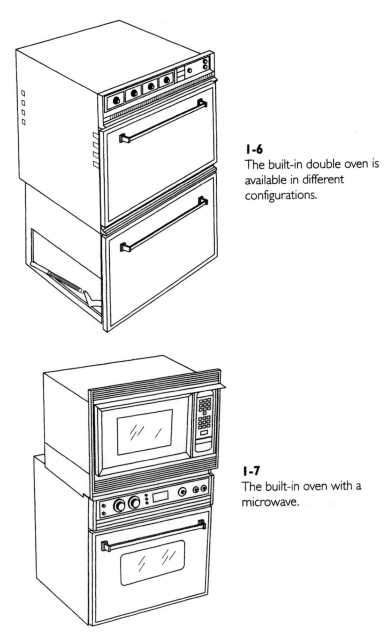

1-6
The built-in double oven is available in different configurations.

1-7
The built-in oven with a microwave.

Listed below are some advantages and disadvantages of built-in ovens (Figs. 1-6 and 1-7):

Advantages
- Can be installed at the most convenient height for the user, putting controls within reach for both a standing or a seated cook.
- Automatic cleaning systems virtually eliminate the task of cleaning the oven manually.

Disadvantages
- Built into the kitchen, and generally cannot be moved when the family moves.
- Installation might involve structural and wiring changes, requiring a carpenter and an electrician.
- Two-oven combinations with microwave ovens often require learning some new cooking techniques.

Eye-level two-oven ranges

Eye-level ranges[3] are 30-inch-wide ranges which have a second oven mounted over the cooktop. The upper oven can be a conventional bake-and-broil oven, or a microwave oven. Designs include features like:
- A selection of either gas burners or electric elements.
- Choice between gas or electric oven.
- Glass cooktop with concealed electric elements underneath.

The lower ovens are available with self-cleaning[1] or continuous-cleaning[2] systems. Microwave oven combinations are available with solid-state microcomputer or electromechanical controls. This type of range gives the maximum cooking capacity in a minimum amount of floor space.

Listed below are some advantages and disadvantages of eye-level one and two-oven ranges (Figs. 1-8 and 1-9):

Advantages
- Controls at eye level might be more convenient for cooks who use walkers, crutches, or a cane.
- Automatic cleaning systems virtually eliminate the task of manually cleaning the oven.
- Can be moved when the family relocates.

Disadvantages
- These controls, and also the oven cavity being located at eye level, are not usable by seated cooks.

CHECKLIST FOR COOKING PRODUCTS

Before going to the store for selecting and purchasing an appliance, read this section on cooking appliances. Fill out the following checklist. Check all that apply.

Measure the area available for the range

_____width
_____depth
_____height

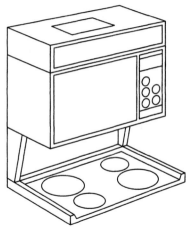

1-8
An eye-level one-oven model, designed to operate on gas or electric.

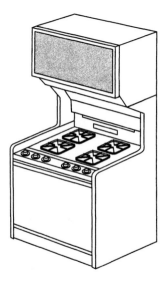

1-9
The eye-level two-oven range (with or without microwave oven combination) is available in a variety of colors.

These measurements are the cut-out measurements, and not the old range measurements. Also, be sure that the range can fit through the doorways of the house.

The type of range desired
☐ Free-standing with one oven
☐ Eye-level (High-low) with two ovens
☐ Eye-level with one oven
☐ Built-in
☐ Drop-in
☐ Slide-in
☐ Electric
☐ Gas
   ☐ LP ☐ Natural

Type of oven needed
- ☐ Single
- ☐ Double
- ☐ Conventional
- ☐ Microwave
- ☐ Combination oven
- ☐ Convection

Oven location
- ☐ Below cooktop
- ☐ One over, one under
- ☐ Separate built-in oven(s)

Oven controls
- ☐ On back console
- ☐ On range front console
- ☐ On hood
- ☐ Dial-type
- ☐ Touch pads
- ☐ Automatic oven clock/timer

Cleaning system
- ☐ Self-cleaning (pyrolytic)
- ☐ Continuous
- ☐ Manual

Broiler type
- ☐ Top-of-oven
- ☐ Low broiler
- ☐ Variable heat

Cooktop style
- ☐ Conventional
- ☐ Glass-ceramic
- ☐ Grille/griddle convertible
- ☐ Induction

Cooktop controls
- ☐ Thermostatic control
- ☐ Eye-level controls
- ☐ Eye-level controls on hood
- ☐ Controls on back splash
- ☐ Controls at front of cooktop
- ☐ Controls at side of cooktop

Accessories
- ☐ Rotisserie
- ☐ Roast temperature probe
- ☐ Griddle

Venting system
- ☐ Separate hood
- ☐ Built-in down draft
- ☐ Vent-microwave oven combination
- ☐ Hood attached to upper oven

☐ Vent over regular cooktop
☐ Vent over grill/griddle cooktop
The preferred color of the range
_____

The people aspect
Who will use the range? _____
How many people are being cooked for? _____
Is the range design convenient for all of the family members? ☐ yes ☐ no
Price range: $_____
Warranty and service information:_____

REFRIGERATORS AND FREEZERS

As with other kitchen appliances, refrigerators and freezers come in a wide variety of styles, sizes, and colors (Fig. 1-10). Some designs might meet the needs of a family member who has a physical or mental limitation better than others.

Some questions to consider when choosing a refrigerator are:

- Does it have a true no-frost system, to do away with manual defrosting?
- Does it have an automatic ice maker, which produces ice without trays to fill or empty?
- Does it have in-the-door dispensers to deliver ice and cold water, without opening the door?
- Does it have shelves, bins, and drawers which pull out to make reachable those foods stored at the back?

There are five basic types of refrigerators that are on the market today.

- Compacts or portables: 1½ to 6 cubic feet (12" to 24" wide)
- Single-door models: 9 to 14 cubic feet (23" to 30" wide)
- Top-mount refrigerator-freezer combination: 10 to 23 cubic feet (24" to 33" wide)
- Bottom-mount refrigerator-freezer: 18 to 20 cubic feet (32"–36" wide)
- Side-by-side refrigerator-freezer: 17 to 30 cubic feet (30"–48" wide)
- Excluding compacts and portables, refrigerators range in height from 56" to 84", and in depth from 24" to 31".

When selecting the capacity of a refrigerator, the following rules should be considered:

- Allow a minimum of 12 cubic feet for the first two persons in the household.
- Add two cubic feet for each additional member. The most popular size for an average family's refrigerator is 17 cubic feet.
- Subtract from this rule if many meals are eaten away from home, using the amount of meals eaten outside the household as a basis.
- Add to the rule if the customer entertains often, if the family is expanding, if there is a vegetable garden growing, or if cooking is enjoyable.

Compact and portables

Compact and portable refrigerators are often used as a supplementary model in family rooms, offices, dorms, vacation homes, campers, and other convenient places.

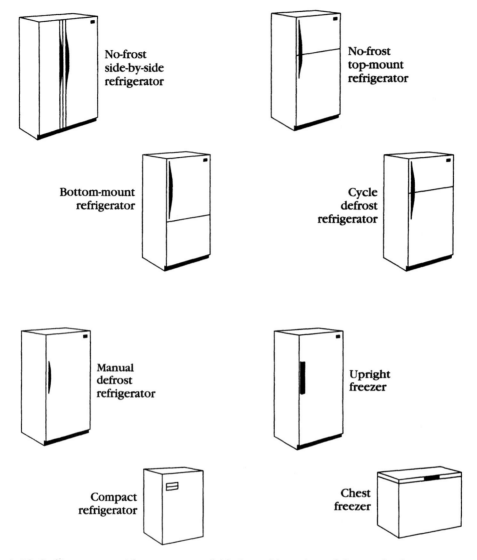

No-frost
side-by-side
refrigerator

No-frost
top-mount
refrigerator

Bottom-mount
refrigerator

Cycle
defrost
refrigerator

Manual
defrost
refrigerator

Upright
freezer

Compact
refrigerator

Chest
freezer

1-10 Refrigerators and freezers are available in a wide variety of sizes and colors.

Many fit on, or under, a countertop. They might be free-standing, or built-in models; and they also come in a variety of colors and finishes.

Listed below are some advantages and disadvantages of compact and portable refrigerators:

Advantages
- Refrigerators of this type have a smaller capacity; less than 6 cubic feet. They can be installed at any height.
- Has a small freezer compartment for ice trays.
- Some models have an optional ice-maker.

Disadvantages
- Frozen food storage is limited to a few days; the maximum a week, depending on the temperature.
- A few compact models have automatic defrosting. However, most are manually defrosted.
- Small size provides limited storage capacity.

Single-door refrigerators

Single-door refrigerators provide both fresh and frozen food storage. Frozen food compartments are located on top of the inside of the refrigerator, and usually contain ice cube trays.

Listed below are some advantages and disadvantages of single-door refrigerators:

Advantages
- Refrigerators of this type generally have a small capacity; less than 14 cubic feet. Most shelf areas are within reach of a seated person.
- Freezer compartments have side opening doors, rather than drop-down doors, for easier accessibility.

Disadvantages
- Refrigerators of this type usually require manual defrosting, a difficult chore for disabled persons.
- Freezer compartments with drop-down doors are inaccessible from a seated position.
- Limited storage capacity.
- Freezer compartment can only be used for short-term storage of commercially frozen food, and for making ice. High sugar foods, such as ice cream, might not stay frozen.

Two-door refrigerators

Two-door refrigerators with top freezers provide fresh and frozen food storage. The freezer maintains a temperature of zero to ten degrees Fahrenheit. These models come with, and without, automatic ice makers.

Listed below are some advantages and disadvantages of two-door refrigerators with top freezers:

Advantages
- Provide proper storage conditions for both fresh and frozen foods. Keeps ice cream frozen.

Disadvantages
- Top freezer is not accessible from a seated position.
- Foods stored near the rear of deep shelves might be difficult to reach without using special aids.

Two-door refrigerators with bottom freezers

Two-door refrigerators with bottom freezers provide both fresh and frozen storage. The freezer maintains a temperature of 0 to 10 degrees Fahrenheit. Models with, and without, automatic ice makers are available.

Listed below are some advantages and disadvantages of two-door refrigerators with bottom freezers:

Advantages
- Freezer shelf and basket slide out for easy accessibility.
- Lower shelves of the fresh food storage area are within easy reach from a seated position.

Disadvantages
- Although this design might meet the needs of some users with disabilities, the bottom-freezer refrigerator is generally not convenient for those in wheelchairs. The low freezer is also inconvenient for users who have trouble stooping or bending.

Side-by-side refrigerators

Side-by-side refrigerators have deep, vertical fresh and frozen food compartments. They require less room, for the doors to open, than other types.

Listed below are some advantages and disadvantages of side-by-side refrigerators:

Advantages
- Provide universal access to the majority of shelves in both refrigerator and freezer compartments.
- Models with in-the-door dispensers give an easy access to ice and ice water without opening the door.
- Models with a third-door, shallow-compartment have the option of putting frequently used foods within easy reach, without opening the main refrigerator door.
- Pull-out shelves, drawers, bins, and baskets provide easier reach for foods stored at the back.

Disadvantages
- Special features add to the total cost.
- Might require more space than available in existing kitchens, due to a wider design.

FREEZERS

A compact, upright freezer will best meet the needs of the disabled person. Installing the freezer on a box or raised platform can help make its contents more easily accessible. Freezers are conveniences for people who do not frequent the supermarket. They are especially useful in homes with smaller refrigerators, and refrigerators having only ice cube tray compartments. Home freezers come in chest and upright models. There are two designs of upright models available on the market today: manual defrost, and automatic defrost. Home freezers are available with wire shelves and baskets, and with storage shelves on the doors in upright models. Listed below are some advantages and disadvantages of freezers:

Advantages
- Make it possible to keep a supply of all kinds of frozen food on hand.
- By stocking up on food at sale prices, and storing them for later use, consumers can easily take advantage of price specials.
- Require fewer shopping trips.

Disadvantages
- Kitchen might not allow space for the freezer.

Checklist for refrigerator and freezer products

Before going to the store for selection and acquisition needs, read and fill out this section on refrigerators and freezers. Check all that apply:

Size
- How large is the space for the freezer?
  _____" Wide, _____" Deep, _____" High
- How much room allows the door to swing open? _____
- Will there be enough room to open the doors completely, so as to remove the storage bins?_____
- Direction of door swing:
  handle on the left side ☐ right side ☐
- How many people will be using the refrigerator?
  ☐ 1 to 2 people (Need at least 12 cubic feet).
  ☐ 3 to 4 people (Need 14 to 16 cubic feet).
  ☐ More than 4 people (Add 2 cubic feet per additional person). To accommodate for later expansion needs, plan for additional refrigerator space, especially if the family is growing larger.
- Plan refrigerator space to accommodate peak loads.
Is there a regular stockade of cold beverages in the refrigerator?
  ☐ Yes ☐ No
- How often does the customer go food-shopping?
  ☐ 2 to 3 times a week
  ☐ Weekly ☐ Daily
- Does the consumer host large holiday dinners?
  ☐ Yes ☐ No
Each "Yes" answer will add to the refrigerator size requirements, as specified above.

Model
- Are there any handicapped or disabled members in the household?
  ☐ Yes ☐ No
A side-by-side model allows easy access to both freezer and fresh food compartments, for those who have limited activity requirements.
- Choose the model desired:
  ☐ Side-by-side
  ☐ Top mount freezer
  ☐ Bottom mount freezer
  ☐ Compact
  ☐ Single door (No long term freezer compartment)
- Which features are important?
  ☐ Automatic defrost
  ☐ Cycle defrost (requires manual defrosting of freezer)
  ☐ Manual defrost (requires manual defrosting of both refrigerator and freezer)
  ☐ Reversible doors
  ☐ Automatic ice maker:
  ☐ Factory installed

☐ Equipped for later installation
☐ Through-the-door dispenser:
    ☐ Ice cubes only
    ☐ Cubes/crushed ice
    ☐ Cubes/crushed ice/chilled water
- Storage drawers:
    ☐ See-through
    ☐ Adjustable temperature
    ☐ Adjustable humidity
    ☐ Sealed snack pack for lunch meats/cheese
- Refrigerator shelves:
    ☐ Glass ☐ Full width only
    ☐ Wire ☐ Half width only
    ☐ Adjustable ☐ Combination full and half width
    ☐ Nonadjustable
- Door storage:
    ☐ Egg compartment
    ☐ Removable
    ☐ Covered
    ☐ Dairy compartment:
    ☐ Butter only
    ☐ Butter and cheese
    ☐ Deep enough for liter size bottles or six packs
    ☐ Removable storage/servers
- Freezer compartment
    ☐ Needed only for 1–2 day storage of frozen food
    ☐ Needed for storage of ice cream, meats, and other frozen food over longer periods
- Freezer shelves:
    ☐ No shelves needed
    ☐ Nonadjustable shelves acceptable
    ☐ Need adjustable shelves
- Convenience features:
    ☐ Juice-can dispenser
    ☐ Ice tray shelves
    ☐ Ice cube bin
    ☐ Wine bottle holder
- Price range: $_____
- Warranty and service information:_____

DISHWASHERS

During the past decades, dishwashers have proven their value and usefulness in reducing the cleanup tasks in the kitchen. They not only save time, energy, and labor, but they also deliver dishes cleaner than those washed by hand.

The most common type is the built-in dishwasher. However, there are other styles available for special situations. When selecting a new dishwasher (Fig. 1-11), there are some questions to consider:

- Is it a front-loading undercounter model? Portables are less convenient because they must be moved into position and hooked to a faucet every time they are used.
- Does it have front controls?
- Does it have a self-cleaning filter, rather than one that has to be removed and cleaned?
- Does it have dispensers and silverware baskets in an easy-to-reach location?
- Are the silverware baskets on the door, or in the bottom rack?
- Are the racks designed with flexibility for easy loading of tall or bulky dishes?

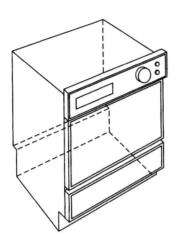

I-II
The undercounter dishwasher is available in 18- or 24-inch width. Portable models are also available.

Built-in models

Built-in models are designed to fit into a 24"-wide space between two kitchen cabinets, and under the countertop. Also available are smaller capacity models that fit into an 18" space.

Convertible/portable models

Convertible/portable models are essentially the same as built-ins, but they have finished sides and tops, drain and fill hoses with a faucet connector, and casters for easy rolling to the sink. These can be installed later as built-ins, if desired.

Undersink model

An undersink model is designed to fit under a special 6-inch deep sink in just 24 inches of space; or under a special double-bowl sink, with a disposer under the second bowl, in 36 inches of space. These dishwashers were designed for small kitchens with limited cabinet space for storage.

Dishwasher-sink combination

A dishwasher-sink combination unit is also available. It includes a stainless steel sink with drainboard, an enameled metal undersink cabinet, and a dishwasher in 48 inches of space. Some "Pullman-type" combination units include a dishwasher as well as a sink, range, and/or, in some, a refrigerator; all in one unit. Listed below are some advantages and disadvantages of dishwashers:

Advantages
- Can save the physical labor of washing dishes.
- Provide out-of-the-way storage for dirty or clean dishes.

Disadvantages
- Require an 18-inch or 24-inch space in the kitchen, so a cabinet might have to be removed.
- Are nearly always designed for use with standard 36-inch high countertops. This might require a two-level counter, if the sink is installed at a 30-inch height for a wheelchair bound person.

Checklist for dishwashers

Before going to the store for selecting and purchasing, ensure that this section on dishwashers has been read and filled out. Check all that apply:

Model
- ☐ Built-in
- ☐ Convertible/Portable
- ☐ Dishwasher-sink Combination
  - Loading convenience
    - ☐ Racks designed to handle large dishes
    - ☐ Adjustable racks
    - ☐ Special baskets:_____
  - Wash system
    - ☐ Water temperature booster
    - ☐ Two or three-level spray arms
    - ☐ Rinse aid dispenser
    - ☐ Self-cleaning filter
    - ☐ Soft food dispenser
  - Control panel
    - ☐ Dials
    - ☐ Pushbuttons
    - ☐ Touch pads
    - ☐ Cycle time indicator
    - ☐ Energy level indicator
    - ☐ System status
  - Cycles
    - ☐ Normal
    - ☐ Rinse only
    - ☐ Economy/water saver
    - ☐ Super-heated rinse
    - ☐ No-heat drying
    - ☐ Light duty
    - ☐ China and glass
    - ☐ Pots and pans
    - ☐ Delay start

- Finishes
    - Tub _____
    - Racks_____
    - Color panels_____
    - Trim kits_____
    - Preferred color_____
- Price range: $_____
- Warranty and service information:_____

LAUNDRY EQUIPMENT

Today's laundry equipment, along with changes in fibers and fashions, usually have eliminated the need for hand laundering, clothesline drying, and routine ironing. There are features for pre-programming the appliance for any type of laundry load. These all give excellent results. From diapers and jeans, to delicate silks and knits, today's laundry system is equipped for all fabric needs.

The typical laundry pair, a standard washer and dryer, will stand side-by-side in 4½ to 5 feet of wall space, depending on the brand and model (Fig. 1-12). Some questions to consider when choosing laundry appliances are:

- Is there enough space available for laundry appliances?
- What control location will be best for the principal user? Some models offer front or rear controls.
- What capacity will best satisfy family needs?
- Will built-in dispensers for bleach and fabric softener increase washer's utility?
- Will the dryer need to be vented outside?
- Which is preferred (gas or electric) for drying clothes?
- How many types of washing cycles are needed?
- How many types of drying cycles are needed?

Automatic washers

Basically, all automatic washers will wash the clothes in the same manner, but there are some key differences in design and special features from model to model, and manufacturer to manufacturer.

1-12
The automatic washer and automatic dryer. The dryer is available in electric or gas.

Top-loading automatic washers

Top-loading models vary in width from about 24 to 30 inches (Fig. 1-12). They are available in a variety of load capacities. Standard capacity washers are built for the average 2 to 4 person household. However, a large capacity model reduces the number of loads washed, saving time. Some models offer front panel controls, and many models have dispensers. Listed below are some advantages and disadvantages of top-loading washers (Fig. 1-13):

Advantages
- Provide a convenient, at-home way to do laundry.
- Models installed in a small space, only 24 inches wide, are available.
- Models with front controls can be reached and operated easily from a seated position.
- Provide a variety of designs and control positions to meet varying user needs.

Disadvantages
- Models with rear console controls are virtually impossible to operate from a seated position.
- Some designs might require special aids to remove loads, set controls, and clean filters from a seated position.
- Compact models, with greatest accessibility from a wheelchair, have a smaller load capacity than other designs.

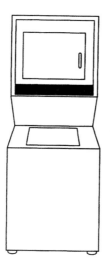

1-13
The top-loading washer with dryer built-in above washer is available in a variety of colors.

Front-loading automatic washers

Front-loading models might have drop-down or side-opening doors (Fig. 1-14). In these models, the entire wash basket revolves. As these clothes tumble, they are lifted by vanes on the side of the basket. Front loaders use less water than top loaders, but they will not handle very large loads because they must have empty space in the drum to tumble clothes. Listed below are some advantages and disadvantages of front-loading washers (Fig. 1-15):

Advantages
- Front controls can be reached and operated easily from a seated position.
- Front opening makes loading and unloading easier for users with limitations.

Disadvantages
- Drop-down door might create wheelchair barrier.
- Door opening might be too low for some wheelchair users, or those who cannot stoop or bend.

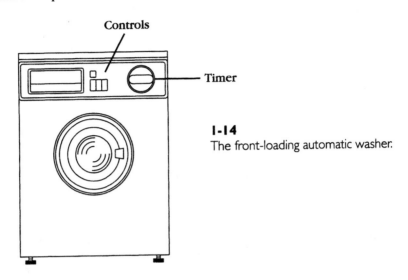

1-14
The front-loading automatic washer.

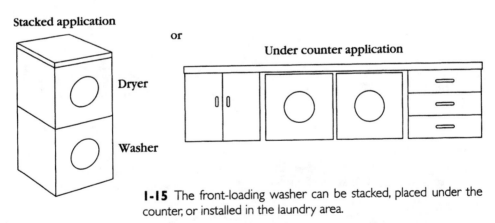

1-15 The front-loading washer can be stacked, placed under the counter, or installed in the laundry area.

Compact automatic washers

Compact, "apartment-sized", washers range from 24" to 27" wide to fit spatial needs. They are available in two forms: built-in, or on casters, so they can be rolled to the kitchen sink for use.

Matching dryers can be installed next to the washer, stacked on a special rack, wall hung, or purchased as a one-piece unit with the washer (Fig. 1-16).

I-16
The compact or portable washer.

Automatic dryers

Automatic dryers perform in the same tumbling manner as front-loading automatic washing machines. But, there are some key differences in design and special features from model to model, and manufacturer to manufacturer.

Dryers are available in a choice of electric or gas. They vary in a variety of load capacities. Some models offer front (or rear) controls, and side-opening (or drop-down) doors. For best efficiency, an electric dryer should have a minimum rating of 4400 watts. Gas dryers require a 120-volt outlet for such features as the motor, lights, and ignition. The gas heater should have a rating of at least 20,000 BTU/hour for top performance. Dryers should be installed in an area that permits proper venting.

Compact, "apartment-sized", dryers range from 24" to 27" wide, so as to fit spatial needs. They are available in two forms: built-in or separate. Compact dryers are electric (either 120-, or 240-volt). The 120-volt dryer takes at least twice as long to dry as the 240-volt model does. While venting is recommended for all dryers, some 120-volt models can be used without venting, if they are not in an enclosed space. The 240-volt dryers must be vented to prevent damage from moisture build-up in the home.

Compact dryers can be installed next to the washer, stacked on a special rack, wall hung, or purchased as a one-piece unit with the washer. Listed below are some advantages and disadvantages of automatic dryers:

Advantages
- Eliminate the difficulties inherent in line drying.
- Give modern fabrics proper care, practically eliminating the ironing chore.
- Designs available to meet the needs of most disabled persons.

Disadvantages
- Models with rear controls are virtually impossible to operate from a seated position.
- Dryer door might be too low without a raised installation.
- As with washers, compact dryers have a smaller load capacity than other designs.

Checklist for washers and dryers

Before going to the store for selecting and purchasing laundry equipment, read and fill out this section on washers and dryers. Check all that apply:

Washer selection chart

- What size is wanted?
    - ☐ Compact/portable
    - ☐ Standard
    - ☐ Large capacity
    - ☐ One-piece washer/dryer
- Style
    - ☐ Front-loading
    - ☐ Top-loading
- Cycle selections
    - ☐ Permanent press
    - ☐ Delicate
    - ☐ Knits
    - ☐ Pre-wash
    - ☐ Soak
    - ☐ Extra-clean
- Options
    - ☐ Variable water level
    - ☐ Water temperature control
    - ☐ Extra rinse cycle
    - ☐ Electronic controls
    - ☐ Water-saver (to reuse wash water)
    - ☐ Small load basket
    - ☐ Bleach dispenser
    - ☐ Fabric softener dispenser
    - ☐ Detergent dispenser
- Color_____ Dryer selection chart
- Which type is preferred?
    - ☐ Electric
    - ☐ Gas
- What size is needed?
    - ☐ One-piece washer/dryer
    - ☐ Standard
    - ☐ Large capacity
- Cycle selections
    - ☐ Permanent press/medium heat
    - ☐ Delicate/low heat
    - ☐ No heat
    - ☐ Timed cycles
    - ☐ Automatic drying
    - ☐ No-heat tumbling at end of the drying cycle
- Other options
    - ☐ Electronic controls
    - ☐ End of cycle signal

☐ Drying shelf
☐ Side-opening door
☐ Drop-down door
- Color_____
- Tub (stainless or porcelain)_____
- Price range: $_____
- Warranty and service information:_____

HOW TO PURCHASE THE BEST APPLIANCE VALUE USING THE ENERGYGUIDE LABEL

Proper planning and evaluation, before buying, can save time, trouble, and money each step of the way. Take the time to determine the appliance's annual cost of operation.

Remember, while some energy-efficient appliances have higher purchase prices than less efficient ones, they will cost less in the long run, because they require less electricity to operate. Calculating and evaluating the appliance's annual cost of operation, is the best way to be prepared for the best buy. Carefully read the EnergyGuide Label, which appears on appliances, to get the best buy.

The EnergyGuide Label (Fig. 1-17) is required by the U.S. government on many home appliances, and gives information to help customers select and save. The information gained from the EnergyGuide Label will be important.

- It will help the customer compare the estimated annual operating costs of one model versus another.
- It will give information about the size range of the models being compared.
- It will tell how each model compares in terms of its energy costs to other models in the same size range.

To read the EnergyGuide Label, at first look for the estimated annual energy cost in the center. To either side are energy costs of lower-rated and higher-rated models. These costs are derived from national-average electricity rates, so knowledge of the local rate for electricity is helpful.

MAJOR APPLIANCE CONSUMER ACTION PANEL (MACAP)

The Major Appliance Consumer Action Panel, or MACAP, is an independent, complaint-mediation group made up of professionals (with expertise in textiles, equipment, consumer law, and engineering) who volunteer their time. Panelists cannot be connected with the appliance industry. They receive no financial remuneration other than the reimbursement of their travel and living expenses while attending meetings. MACAP receives comments and complaints from appliance owners; excessive charges, delays in parts and service, and alleged unnecessary repairs are frequently mentioned. Other complaints are related to product performance; such as operating noise, temperature maintenance, and running time. Non-responsiveness of dealers and manufacturers, warranty coverage, food loss claims, imperfect finishes, improper installation, and purchasing dissatisfaction are also alleged. MACAP also studies industry practices, and advises industries of ways to improve their services to consumers. It also recommends how to educate consumers on proper appliance purchase, use, and care. The panel develops and distributes educational

ENERGYGUIDE

Estimates on the scale are based on a national average electric rate of 7.70¢ per kilowatt hour and a natural gas rate of 55.20¢ per therm.

Only standard size dishwashers are used in the scale.

Electric Water Heater

Model with lowest energy cost
$71
Model with highest energy cost

$39
$78

▼ THIS MODEL ▼ ▼
Estimated yearly energy cost

Gas Water Heater

Model with lowest energy cost
$32
Model with highest energy cost

$22
$41

▼ THIS MODEL ▼ ▼
Estimated yearly energy cost

Your cost will vary depending on your local energy rate and how you use the product. This energy cost is based on U.S. Government standard tests.

How much will this model cost you to run yearly?

with an electric water heater

Loads of dishes per week		2	4	6	8	12
Estimated yearly $ cost shown below						
Cost per kilowatt hour	2¢	$6	$12	$18	$24	$36
	4¢	$12	$24	$36	$48	$71
	6¢	$18	$36	$54	$71	$107
	8¢	$24	$48	$71	$95	$143
	10¢	$30	$60	$89	$119	$179
	12¢	$36	$71	$107	$143	$214

with a gas water heater

Loads of dishes per week		2	4	6	8	12
Estimated yearly $ cost shown below						
Cost per therm (100 cubic feet)	10¢	$5	$10	$15	$21	$31
	20¢	$6	$13	$19	$25	$38
	30¢	$7	$15	$22	$30	$45
	40¢	$9	$17	$26	$34	$52
	50¢	$10	$19	$29	$39	$58
	60¢	$11	$22	$33	$44	$65

Ask your salesperson or local utility for the energy rate (cost per kilowatt hour or therm) in your area, and for estimated costs if you have a propane or oil water heater.

Important Removal of this label before consumer purchaser is a violation of federal law (42 U.S.C. 6302)

1-17 The EnergyGuide Label that appears on most appliances.

publications and periodic news releases, when its review of individual consumer complaints pinpoint information that would be useful to consumers. The panel is sponsored by the Association of Home Appliance Manufacturers.

The types of appliances represented are:

- Compactors, and laundry equipment
- Dehumidifiers, ranges, and microwave ovens
- Dishwashers, refrigerators, and freezers
- Disposers, and room air conditioners

After a complaint reaches MACAP, the staff screens it, so as to ascertain whether the consumer has already requested assistance from the local dealer and brand-name owner's headquarters office. If this has been done, the complaint is sent to the sponsoring association and, thereby, enters MACAP's communications phase. Here:

1. Copies of the consumer's correspondence, or a summary of telephone comments, are sent to the brand-name owner involved, requesting a report on proposed action within two weeks.
2. A letter is sent to the consumer acknowledging receipt of the complaint, reporting action taken, and asking for any additional information, if necessary.
3. When an answer is received from the brand-name owner, the staff writes a letter to the consumer confirming any company action or information. The consumer is asked to return a card verifying this action.
4. The file is then resolved, unless the consumer otherwise advises MACAP.
5. If the complaint reaches an impasse, the file moves to the study phase.

In the study phase, the Panel discusses the file at a meeting. Preparation for this review includes the gathering of an exact and detailed background of the complaint from the consumer, and from the brand owner, if such information is not already on file. If conflicting reports are received, the Panel might ask for an independent, on-site evaluation by a utility, extension, university home economist, or an engineer. All information is included in a confidential summary prepared by the staff and discussed by MACAP. The panel might make a recommendation to the company and/or consumer, ask for additional information, or close the file on the basis of the information presented.

Purchasing decisions

The purchase of a major appliance is one of the most important investments made for the home. MACAP experience in handling consumer appliance complaints has shown that poor purchasing decisions lead to unhappy appliance owners.

Here are some pointers from MACAP to help the consumer make wise decisions:

- Ask the dealer for specification sheets from several manufacturers of the appliance types you plan to purchase. Study them carefully, and note the different features, designs, and capacities.
- Ask the dealer to see the warranty before purchasing the appliance. Does the warranty cover the entire product? Only certain parts? Is labor included? How long is the warranty coverage?
- Ask the dealer for the use and care manual. Read it carefully before you purchase the appliance. The dealer should have manuals available from the floor models on display. These manuals will help in asking pertinent questions; and it will explain how the product operates, and what special care it needs.
- Decide what special features are essential. Consider the possibility of adding on features at a later date, such as an icemaker for a refrigerator.
- Check the space available for the appliance. Will it fit where it is planned? Is there adequate clearance space in the hallways and doors through which the appliance will have to pass before installation?

- Check the product design carefully prior to purchase. Compare the designs of different brands. If a combination microwave oven/range is being purchased, check the space between the units, to be sure everything will fit.
- Clearly establish the cost of delivery and installation. Are these costs included, or are they extra?
- Ask the dealer if he services the appliances he sells. If not, ask him where to go for authorized factory service.
- Compare price in relation to convenience and service. Both vary according to the model. As more features and conveniences are included, the price increases.
- Be sure the house has adequate electrical service for the appliance, in order to avoid overloading circuits. Also, be sure it has adequately grounded, three-hole receptacles.[4]

APPLIANCE WARRANTIES

MACAP urges the consumer to compare warranties of different brands the way you compare price, size and features. It's a basic consumer responsibility! Yet the panel's experience in working with consumer appliance problems indicates many consumers are not aware of the importance of warranty buymanship.

Warranty inspection is a legislated consumer right. A federal law, the Magnuson-Moss Warranty/Federal Trade Commission Improvement Act of 1975, requires warranty information to be available at the point of purchase for products costing $15 or more. The law does not, however, require manufacturers to provide warranties on their products.

The store must provide this information in one of the following ways:
- Displayed near the appliance.
- Shown on the package displayed with the appliance.
- In an indexed and updated binder, that is prominently displayed, containing all warranties for products sold in the department.
- If ordering the appliance through a catalog, the catalog must include the warranty text, or offer it upon request.[5]

"Full" vs. "limited" warranties

The "full" warranty offers more protection. Under a "full" warranty, as a minimum, the warrantor must remedy the problem within a reasonable time and without charge, for as long as the warranty is in effect. In addition, the manufacturer might not limit the duration of any implied warranty. Any limitation (or exclusion) of consequential damages (e.g., food loss, or floor damage) must appear conspicuously on the warranty. If a reasonable number of attempts to correct an in-warranty problem fail, the consumer must be given the choice of a replacement, or a refund.

Under a "limited" warranty, the protection is limited to what is outlined in the terms of the warranty. For example, the warranty might require that the consumer pay for diagnostic costs, labor costs, and other costs of that kind. Also, any implied warranty might be limited to the duration of the written warranty. A "limited" warranty should be studied carefully to determine exactly what the warrantor will provide, and what costs the consumer must pay.[5]

Warranty time limitations

Some major appliances are hardly used over a year's time. The window air conditioner in Minnesota, the refrigerator at the lake cabin in Michigan, the range in a Florida vacation condominium; all of these have only seasonal use.

When problems with these appliances arise, some owners feel that four years of seasonal usage should constitute less than one year's worth of warranty coverage. The manufacturers think differently. MACAP knows it is necessary for the consumer to be aware of the contents of their appliance warranties, and also for how long the period covers. The coverage is stated for a specific period of time, not for how often the appliance is used.

The coverage period might vary with the product's components. For example, a window air conditioner warranty might provide free repairs of any part that breaks down in the first year, but only partially cover repair expenses of the sealed system components (compressor, evaporator, etc.) for an extra four years. Manufacturers can offer almost any type of warranty, as long as the provisions are clearly stated, and the warranty is available for review before buying.

Although seasonal usage might result in less wear and tear on the appliance, MACAP believes that there are at least two valid reasons for not basing warranties on usage:

- Basing a warranty on usage (rather than appliance age) is unrealistic, because it's impossible, for a manufacturer to monitor actual consumer usage.
- Some appliances actually suffer more from extended periods of non-use, than from continuous daily operation.

For example, refrigerator and air conditioner sealed systems stay vacuum tight (generally for many years) if the refrigerant gas and oil it contains are kept moving. This way, the various seals are kept lubricated and soft. They might dry out if the mechanisms involved don't move for long periods. The dishwasher provides yet another example. Many consumers don't realize that the seals in the pump area of the dishwasher are meant to be immersed in water at all times. Water keeps the seal soft and pliable, which is necessary for proper operation.

MACAP urges consumers to carefully read the warranty that comes with each major appliance, and use the appliance enough during the warranty period so that any defects might have time to surface. It is important to use every feature and control on the new appliance soon after it is installed.[6]

APPLIANCE INSTALLATION INSTRUCTIONS

The manufacturer supplies the installation instructions with every new appliance that is purchased. These instructions will help the installer to plan, locate, install, and secure the appliance for proper operation.

Appliance use and care manual

These manuals contain information and suggestions to help the customer get the best results from their appliances. The manual will disclose to the customer how to start their appliance, maintain it, and how to use all the features that come with their appliance. Also, included in the use and care manual are the following:

- Safety precautions
- Parts and features
- What to do before using the appliance
- How to use the appliance
- Maintenance instructions
- Common problems and solutions
- Vacation and moving care
- Assistance
- Warranty information

WHERE TO GET HELP IF NEEDED

Keep careful records. Always put complaints in writing, and keep copies of all correspondence and service receipts. Be sure to ask for service receipts, even for no-charge, in-warranty calls. Note details. When the problem was first noticed, when it was reported, and the servicing history (who serviced the appliance, when, what was done, and how often service was required).

If there are complaints about the appliance, there are three steps to follow:
- Read the use and care manual that comes with the appliance. Also, check the plug, as well as fuses, pilots, and controls.
- Call the service company authorized to fix the brand. They have the training and equipment to deal with appliance service problems.
- Then, if not satisfied, contact the manufacturer's main customer relations office. This address and phone number is located in the use and care manual.

If none of these steps solve the problem, then write to:

Major Appliance Consumer Action Panel
20 North Wacker Drive
Chicago, IL 60606

Endnotes

1. Pyrolytic cleaning is the true self-cleaning system. It uses high heat during a special 1 to 3 hour cycle to decompose food, soil, and grease. During the cycle, which is clock-controlled, the oven door is latched and locked. It cannot be opened until the oven cools down. All of the oven walls, racks, and the door (except for a small area outside the door gasket) are completely cleaned. After cleaning, a small bit of white ash might be found, which can be easily wiped up.
2. Catalytic and continuous cleaning ovens use a special porous coating on the oven walls that partially absorbs and disperses the soil. This process takes place during normal baking and keeps the oven presentably clean, but the racks and door parts must be cleaned by hand. Some manufacturers recommend occasionally operating an empty oven at 500 degrees Fahrenheit, to remove any build-up of soil. This special oven coating cannot be cleaned with soap, detergent, or commercial oven cleaners without causing permanent damage.

3. Eye-level ranges are also referred to as high-low ranges, and also as tri-level ranges.
4. Reprinted from MACAP Consumer Bulletin, issue no. 8, December, 1985.
5. Reprinted from MACAP Consumer Bulletin, issue no. 1, December, 1979.
6. Reprinted from MACAP Consumer Bulletin, issue no. 5, February, 1983.

2
Safety precautions

Safety starts with accident prevention. Injuries are usually caused because learned safety precautions are not practiced. In this chapter are listed some tips to help the technician to correctly and safely install, operate, and repair major appliances.

Any person who cannot use basic tools should NOT attempt to install, maintain, or repair any major appliance. Any improper installation, preventative maintenance, or repairs will create a risk of personal injury, as well as property damage. Call the service manager if installation, preventative maintenance, or the repair procedure is not fully understood.

Every technician should carry a first aid kit. They should know how to properly use the contents of that kit, and know its location. Technicians should carry a fire extinguisher in their service vehicles, in case of an emergency. It is also recommended that they take a first aid course, such as those offered by the American Red Cross.

SAFETY PROCEDURES

Individual, electrical, chemical, appliance, operating, and installation safety precautions are generally the same for all major appliances. Carefully observe all safety cautions and warnings that are posted on the appliance being worked on. Understanding and following these safety tips can prevent accidents.

Individual safety precautions

Protecting yourself from injuries is necessary. Before installing, maintaining, or servicing any major appliance, do the following:

- Wear gloves. Sharp edges on appliances hurt hands.
- Wear safety shoes. Accidents are often caused when dropping heavy appliances, especially on feet that are not protected.
- Avoid loose clothing that could get caught in the appliance while it is operating.
- Remove all jewelry when working on appliances.
- Tie long hair back.
- Wear safety glasses to protect your eyes from flying debris.
- Use proper tools, in clean and good condition, when repairing appliances.
- Have ample light in the work area.
- Be careful when handling access panels, or any other components that may have sharp edges.

- Avoid placing hands in any area of the appliance that have not been visually inspected for sharp edges, or pointed screws.
- Be sure that the work area is clean and dry from water and oils.
- When working with others, always communicate with each other.
- Always ask for help to move heavy objects.
- When lifting heavy appliances, always use your leg muscles and not your back muscles.

Electrical safety precautions

Know where, and how, to turn off the electricity to the appliance. For example: plugs, fuses, circuit breakers, and cartridge fuses; know their location in the home. Label them. If a specific diagnostic check requires that voltage be applied, reconnect electricity only for the time required for such a check, and disconnect it immediately thereafter. During any such check, be sure no other conductive parts come in contact with any exposed current-carrying metal parts. When replacing electrical parts, or reassembling the appliance, always reinstall the wires back on their proper terminals according to the wiring diagram. Then check to be sure that the wires are not crossing any sharp areas, nor pinched in some way, nor between panels, nor between moving parts that may cause an electrical problem. These additional safety tips are also important to remember:
- Always use a separate, grounded electrical circuit for each major appliance.
- Never use an extension cord for major appliances.
- Be sure that the electricity is off before working on the appliance.
- Never remove the ground wire from a three-prong power cord, or any other ground wires from the appliance.
- Never bypass or alter any appliance switch, component, or feature.
- Replace any damaged, pinched, or frayed wiring before repairing the appliance.
- Be sure all electrical connections within the product are correctly and securely connected.

Chemical safety precautions

Chemicals are also dangerous. Knowledge of chemical safety precautions is essential at all times. The following tips are examples of important practices:
- Remove all hazardous materials from the work area.
- Always store hazardous materials in a safe place, and out of the reach of children.
- Never smoke, or light a flame, when working on gas appliances.
- Before turning on appliances that use water, run all of the hot water taps in the house for approximately five minutes. This clears out the hydrogen gas that can build up in the water heater and pipes, if not used for more than two weeks.

Appliance safety

Call the service manager to check out the appliance, if the safety of an appliance is in doubt.

Only use replacement parts of the same specifications, size, and capacity as the original part. If you have any questions, contact your local appliance parts dealer, or your service manager.

Check water connections for possible water leaks before reconnecting the power supply. Then, completely reassemble the appliance, remembering to include all access panels.

Operating safety

After repairing the appliance, do not attempt to operate it unless it has been properly reinstalled, according to the use and care manual, and the installation instructions supplied by the manufacturer. If these instructions are not available, do not operate the appliance. Call the service manager to check out the reinstallation, or ask for a copy of the installation instructions from the manufacturer.

Know where the water shutoff valves are located for the washer, dishwasher, ice maker, and water heater, as well as the house's main water shutoff valve. Label them. Following these additional safety tips can also prevent injuries:

- Do not allow children to play on, or to operate, appliances.
- Never allow anyone to operate an appliance if they are not familiar with its proper operation.
- When discarding an old appliance, remove all doors, to prevent accidental entrapment and suffocation.
- Instruct the customer to only use the appliance for the job that it was designed to do.

Installation safety precautions

The first step in assuring safety with major appliances is to be sure that they are installed correctly. Be sure to read the installation instructions, and the use and care manual, that comes with the appliance. Observe all local codes and ordinances for electrical and plumbing connections. Ask your local government agency about these codes. Additional safety tips:

- Carefully observe all safety warnings that are contained in the installation instructions, and in the use and care manual.
- The work area should be clear of unnecessary materials, so that there is plenty of room to work on the appliance.
- Be sure that the appliances are installed and leveled on a floor strong enough to support their weight.
- The appliance should be protected from the weather, and from freezing, or overheating.
- The appliance should be correctly connected to its electric, water, gas, drain, and/or exhaust systems. It should also be electrically grounded.
- Be sure that the appliance has a properly installed anti-tip device, as in the case of kitchen ranges.

GROUNDING OF APPLIANCES

In 1913, the National Electrical Code (NEC) made grounding, at the consumers home, mandatory. The NEC required that range frames be grounded to a neutral conductor in 1943. Later, in 1946, it required that receptacles in laundry areas be grounded. Soon after, the NEC required that the frames of automatic dryers be grounded to neutral conductors. Then, in 1959, NEC required that automatic washers, automatic dryers, automatic dishwashers, and certain motor-operated hand-held appliances to be grounded. All 15 amp and 20 amp branch circuits have grounding type receptacles, as specified in 1962. Finally, in 1968, the code required that refrigerators, freezers, and air conditioners be grounded.

The greatest importance of grounding appliances is because it prevents people from receiving shocks from them. However, the major problem associated with the adequate grounding of appliances is that many homes are not equipped with three-prong grounded receptacles. To solve this problem, the consumer must install, or have installed, a properly grounded and polarized three-prong receptacle. A qualified electrician should connect the wiring, and properly ground and polarize the receptacle.

Remember that safety is the paramount concern, especially when dealing with electricity. Both the technician and the consumer must be aware that it only takes about one hundred milliamperes of current to cause death in one second. Here are some safety tips:

- Do not install or operate an appliance unless it is properly grounded.
- Do not cut off the grounding prong from the appliance plug.
- Where a two-prong wall receptacle is encountered, it must be replaced with a properly grounded and polarized three-prong receptacle.
- Call the service manager if you doubt your abilities. When dealing with electricity, there is no leeway for mistakes.

CHECKING APPLIANCE VOLTAGE

If it becomes necessary to test an appliance with the voltage turned on, observe the following precautions:

- The floor around the appliance must be dry. Water and dampness increase the probability of a shock hazard.
- When using a multimeter, always set the meter correctly for the voltage being checked.
- Handle only the insulated parts of the meter probes.
- Touch components, terminals, or wires with the meter probe tip only.
- Touch the meter probe tips only to the terminals being checked. Touching other components could damage good parts.
- Be sure that the appliances have properly installed anti-tip devices, as found on kitchen ranges.

3
Tools needed for installation and repair

A basic knowledge of hand tools, electrical tools, and test meters is necessary to effectively complete most installations and repairs. This chapter will cover the basics of each tool. A working knowledge of these tools is a must for the installation and repair of appliances. Always follow safety precautions and manufacturer's recommendations when handling tools.

Before starting on any type of repairs, take the time to put together a toolkit with a selection of good quality hand tools. A partial list of common hand tools includes:

Screwdrivers a complete set of flat-blades ranging from ⅛ inch to ⁵⁄₁₆ inch. Handle sizes may vary with the blade dimension. Phillips-tip sizes also vary; the two most common are #1 and #2.

Nut drivers a complete set is recommended. The common sizes are: ³⁄₁₆, ¼, ⁵⁄₁₆, ¹¹⁄₃₂, ⅜, and ½ inch.

Wrenches
- Socket wrenches, either 6-point or 12-point ranging in sizes from ⁵⁄₃₂ to 1 inch.
- Box wrenches; common sizes range from ¼ to 1½ inch.
- Open-end wrenches; common sizes range from ¼ to 1⅝ inch.
- Adjustable wrenches. The handle size indicates the general capacity. For example, a 4-inch size will take up to a ½-inch nut. A 16-inch handle will take up to a 1⅝-inch nut.
- Allen wrenches, have sizes ranging from ¹⁄₁₆ to about ½ inch.
  ~ Claw hammer
  ~ Adjustable pliers
  ~ Flashlight
  ~ Drop-cloth

SAFETY PRECAUTIONS

Safety starts with accident prevention. Listed in this chapter are some tips to help the technician when using hand and power tools.

Warning: Any person who cannot use basic tools should *not* attempt to install, maintain, or repair any major appliance. Any improper installation, preventative maintenance, or repairs creates a risk of personal injury and property damage.

Individual safety precautions

Injuries abound when using tools. To be protected from injuries when using hand tools and power tools, do the following:
- Wear gloves.
- Avoid wearing loose clothing when working with power tools.
- Wear safety glasses to protect the eyes from flying debris.
- Use tools according to manufacturers specifications, and never alter their use.

Safety precautions when handling tools

Regardless of which tool is being used, these same rules of care and safety apply:
- Keep tools clean and in good working order.
- Use the tool only for jobs for which it was designed.
- When using power tools, be certain that the power cord is kept away from the working end of the tool.
- If the tool has a shield or guard, be sure it is working properly, and remember to use it.
- If an extension cord is used, be sure it is in good working order. Do not use it, if there are bare wires showing. Also, use a heavy-gauge wire extension cord to ensure adequate voltage for the tool being used.
- Be sure that the extension cord is properly grounded.
- Grip the tool firmly.
- Never use worn-out tools. A worn-out tool has more potential for causing injuries. For example, with a worn-out screwdriver, there is a greater possibility for slips, which would make medical attention necessary.
- If there is a problem with a power tool, never stick your fingers in the tool. Unplug it first, and then correct the tool's problem.

SCREWDRIVERS

A screwdriver is a hand tool used to attach, or remove, screws. The two most common types are the flat-blade and Phillips. The flat-blade screwdriver is used on screws that have a slot in the screw head (Fig. 3-1). The flat-blade screwdriver is available in many sizes and shapes. Always use the largest blade size that fits snugly into the slot on the screw head, so that it will not slip off the screw. The screwdriver should never be used as a pry bar, or a chisel: it was not designed for that purpose.

A Phillips screwdriver is a hand tool that is used to attach, or remove, screws that have two slots crossing at right angles in the center of the screw head (Fig. 3-2). When using a Phillips screwdriver, exert more pressure downward in order to keep the tool in the slots. Always use the largest Phillips size that fits snugly into the slots, just as with the flat-blade.

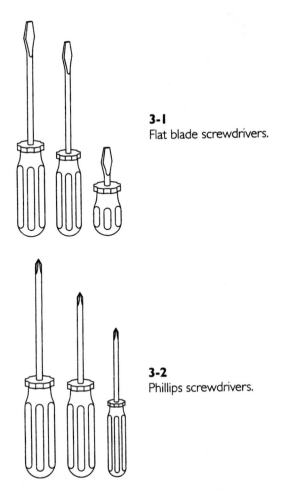

3-1
Flat blade screwdrivers.

3-2
Phillips screwdrivers.

Never use worn screwdrivers when working on appliances; because a worn screwdriver may damage the head of the screw. It can also damage the product on which you are working.

NUT DRIVERS

Many manufacturers use metal screws with a hexagonal head. A nut driver is a hand tool similar to a screwdriver, except that the working end of the driver is hexagonal-shaped, and fits over a hexagonal nut or a hexagonal bolt head. Each size nut requires a different sized driver (Fig. 3-3).

WRENCHES

Wrenches are the most frequently used tool. There are many types and sizes of wrenches. Their purpose is to turn nuts and bolts (Fig. 3-4). Wrenches are generally available in five different types: socket wrenches, box wrenches, open-end wrenches, adjustable wrenches, and Allen, or hex, wrenches.

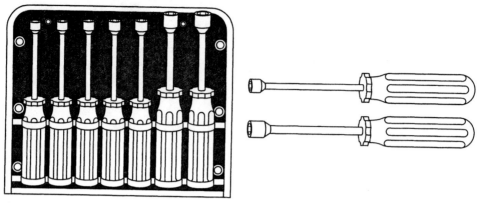

3-3 Hex-nut drivers.

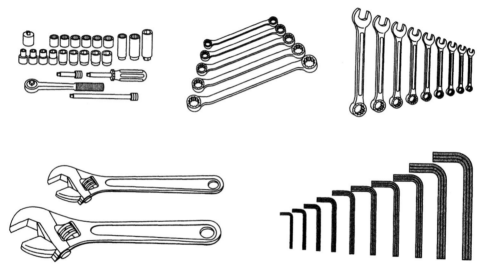

3-4 Wrenches are used to remove and to fasten nuts and bolts. They are available in socket wrench, box, open-end, adjustable, and Allen types.

Socket wrenches are used to slip over bolt heads, as opposed to other wrenches listed that are used at right angles to the nut or bolt. This arrangement allows more leverage to be applied to loosen, or tighten, the nut or bolt (Fig. 3-5). Always choose the correct wrench for the job, using box wrenches for heavy-duty jobs. Open-ended wrenches are useful for medium-duty work. Adjustable wrenches help light-duty jobs, and also odd sized nuts and bolts.

Box wrenches are useful in certain close-quarter situations. Open-end wrenches are needed where it is impossible to fit a socket, or a box wrench, on a nut, bolt, or fitting from the top. The adjustable wrench is useful where a regular open-end wrench could be used. It is adjustable to fit any size object within its maximum opening.

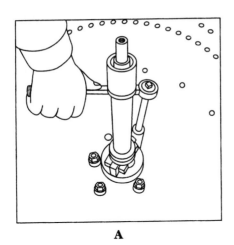

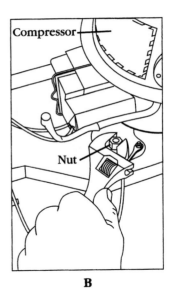

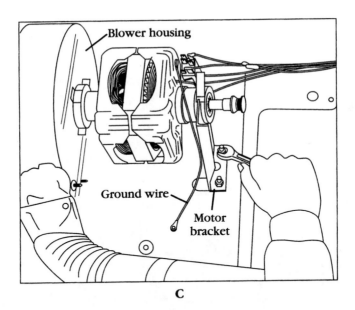

3-5 A. Socket wrench. B. Adjustable wrench. C. Open-end wrench.

Allen wrenches are used for adjusting and removing fan blades, or other components that are held in place by Allen set screws. The Allen wrench has a 6-pointed flat face on either end.

HAMMERS

A hammer is a hitting tool. There are many sizes and styles of hammers made (Fig. 3-6). The most common type used in appliance repairs is the claw hammer. The claw hammer can also be used for prying objects.

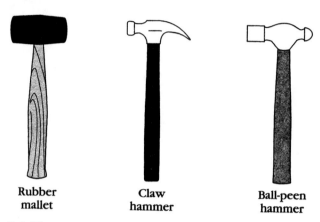

| Rubber
mallet | Claw
hammer | Ball-peen
hammer |

3-6 The most commonly used hammer is the claw hammer.

PRYING TOOLS

Prying tools are available in many sizes and shapes. The most common are the crowbar, ripping bar, and the claw hammer. The claw hammer is basically used for light duty work: removing nails and prying small objects. The ripping bar is used for medium-grade work, and the crowbar for the heavier work.

PLIERS

Pliers are one of the most frequently used tools. A pliers is a tool for holding or cutting, depending on the type. Generally, they are not made to tighten, or unscrew, heavy nuts, and bolts. They are available in many sizes and shapes (Fig. 3-7). Choose a pliers to fill a particular need, being careful that it is the proper pliers for the job. These are some of the most common types of pliers:
- Slip-joint
- Slip-joint adjustable
- Vise grip
- Needle nose
- Diagonal cutting

Slip-joint pliers are pliers for everyday tasks (Fig. 3-7A). The jaws can be adjusted into two different positions. Do not use them on nuts, bolts, or fittings. They can easily slip and injure both the technician and the device.

Slip-joint adjustable pliers are also used for general jobs (Fig. 3-7I). They would be preferred over slip-joint pliers when working on a larger object. The jaws of slip-joint adjustable pliers can be moved into many different positions.

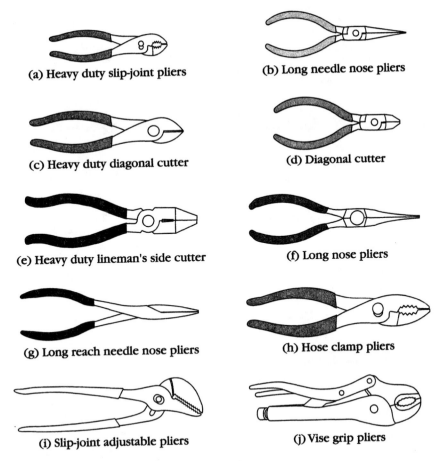

(a) Heavy duty slip-joint pliers

(b) Long needle nose pliers

(c) Heavy duty diagonal cutter

(d) Diagonal cutter

(e) Heavy duty lineman's side cutter

(f) Long nose pliers

(g) Long reach needle nose pliers

(h) Hose clamp pliers

(i) Slip-joint adjustable pliers

(j) Vise grip pliers

3-7 Pliers are available in many sizes and shapes.

The *vise grip pliers* are actually four tools in one (Fig. 3-7J): a clamp, a pipe wrench, a hand vise, and pliers. The lever holds the jaws in one position, allowing the vise to hold up to one ton of pressure.

The *needle nose pliers* are mostly used with electronic, telephone, and electrical work (Figs. 3-7B, F, and G). Other uses would be use in confined areas, to form wire-loops, and to grip tiny pieces firmly. This is because of the long nose. They are also available with side cutters.

Diagonal cutting pliers are used in electrical and electronic work (Fig. 3-7C, D, E). They are used for cutting wire and rope.

CUTTING TOOLS

There are many different types of tools used for cutting. The object is to know which tool to use in each situation.

Chisels are used for cutting metal and wood. They are made of high carbon steel, to make them hard enough to carve through metal (Fig. 3-8). These should be used when removing rusted bolts and nuts.

Cutting tools   **41**

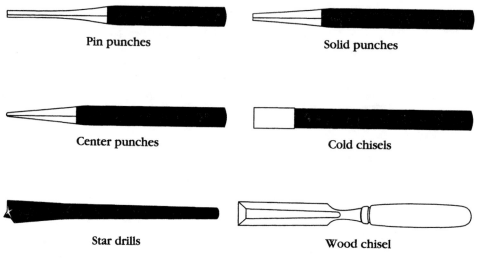

Pin punches

Solid punches

Center punches

Cold chisels

Star drills

Wood chisel

3-8 Chisels are available in many different sizes and shapes, such as wood chisels, metal chisels, and concrete chisels.

Hacksaws are used for cutting metal (Fig. 3-9). The hacksaw consists of a handle, frame, and a blade. The frame is adjustable, so it can accept any length of blade. The blades are available with different numbers of teeth per inch.

A file is a cutting tool. It is used to remove excess material from objects. They also come in a variety of sizes and shapes (Fig. 3-10).

Drill bits are also cutting tools. They are designed for cutting holes in metal, wood, and concrete (Fig. 3-11).

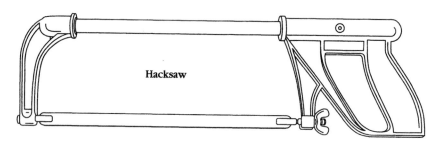

Hacksaw

3-9 Saws are used to cut metal, wood, etc.

POWER TOOLS

Power tools do the same job as hand tools. However, they do the job faster (Fig. 3-12). The most common power sources can be either electric or battery power. When using power tools, there are some safety precautions that must be followed to prevent accidental injury. Always read the use and care manual that comes with each power tool.

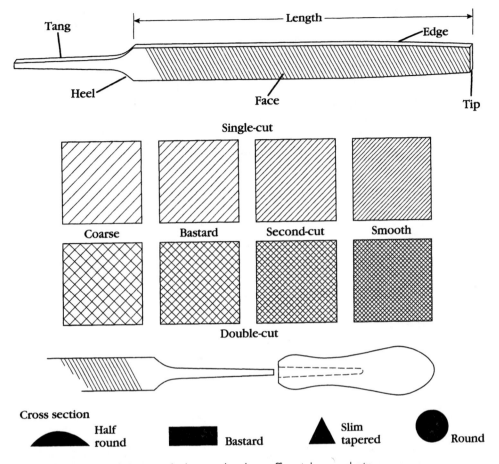

Single-cut

Coarse   Bastard   Second-cut   Smooth

Double-cut

Cross section

Half round   Bastard   Slim tapered   Round

3-10 Files are used to smooth the rough edges off metals, wood, etc.

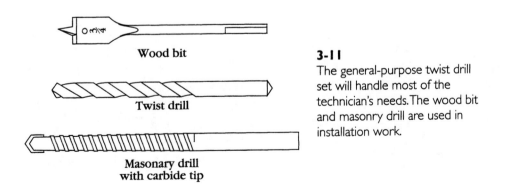

Wood bit

Twist drill

Masonary drill
with carbide tip

3-11
The general-purpose twist drill
set will handle most of the
technician's needs. The wood bit
and masonry drill are used in
installation work.

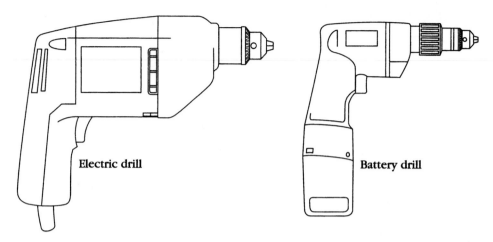

Electric drill   Battery drill

3-12 Power tools can have either a line- or battery-operated power supply.

SPECIALTY TOOLS

These tools are specifically designed for a particular use, and are used for in-depth servicing of the appliance. For example, these type of tools are used for the installation and removal of special screws and nuts; they are required to remove the bearings in washing machines; and, also for adjusting switch contacts. Figure 3-13 illustrates some of the many types of specialty tools. Other specialty tools, and their uses, will be mentioned in the later chapters of this book.

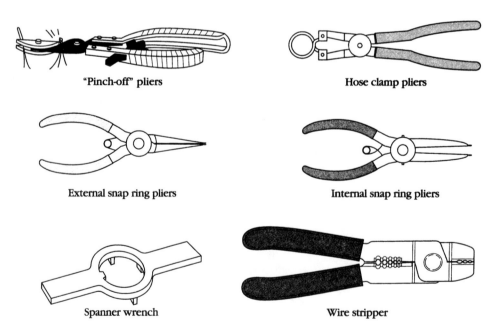

"Pinch-off" pliers   Hose clamp pliers

External snap ring pliers   Internal snap ring pliers

Spanner wrench   Wire stripper

3-13 In addition to the basic hand tools, a number of special-purpose tools used in appliance servicing are illustrated.

44 Tools needed for installation and repair

TEST METERS

Test instruments are very important tools, and they are used in assisting a diagnosis of the various problems that arise with appliances (Fig. 3-14). Listed below are some varieties of test meters:

- The volt-ohm-milliammeter is a test instrument used for testing the resistance, current, and voltage of the appliance. It is also the most important test meter to have in the tool box.
- An ammeter is a test instrument connected into a circuit to measure the current of the circuit, without interrupting the electrical current.
- A wattmeter is a test instrument used to check the total wattage drawn by an appliance.
- A temperature tester is a test instrument used to measure the operating temperatures of the appliance.

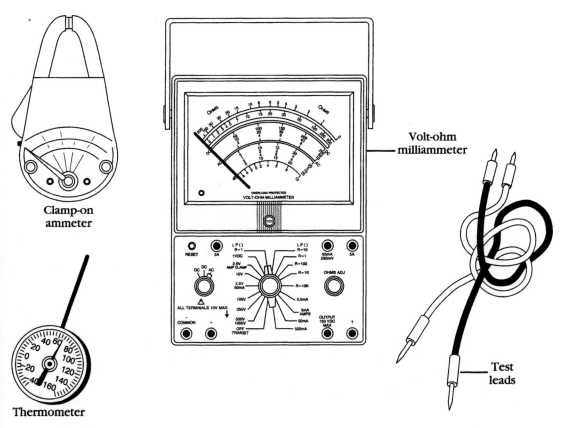

3-14 Test meters are available with analog or digital readouts. Thermometers display the temperature.

4
Basic techniques

There are many different ways to diagnose a problem, but all of them basically use the same reasoning of deduction.

- Where the consumer thinks the malfunction is located within the appliance.
- Where the actual problem is located within the appliance.
- Are there any related problems with the appliance?
- How to solve the problem with the appliance.

For example, the consumer states that the dryer does not dry the clothes, and believes that the heating element is bad. The actual problem might be a restricted exhaust vent, a clogged lint filter, bad heating elements, faulty operating thermostats or safety thermostat, or improper control settings.

When checking the dryer, you notice that the control settings are set for air-drying, instead of heat-drying. Thus, the actual problem was the control settings not being positioned correctly. The related problem is "How did the control setting move to the air dry position?" And then, the question of "Does the consumer know how to operate the dryer?" To solve this problem, you will have to instruct the consumer in the proper operation of the dryer.

All appliances go through a certain sequence of events. Understanding the proper operation, and this sequence as indicated in the use and care manual, is beneficial when diagnosing the appliance.

PRODUCT FAILURE

Given the information about the appliance's problem, information and diagnostic charts from this book, and the information you have read in the use and care manual; as the servicer, you will be able to perform the following steps in sequence to diagnose and correct a malfunction. Listed below are the basic steps to follow when diagnosing an appliance problem.

1. *Verify the complaint* Ask the consumer what symptoms were caused by the problem with the appliance.
2. *Check for external factors* For example: is the appliance installed properly; does the appliance have the correct voltage; etc.?
3. *Physical damage* Check for internal and external physical damage. Any damage will prevent the appliance from functioning properly. Two examples are: broken parts at the base of the washing machine, or a damaged cabinet that will prevent the doors from closing properly.
4. *Check the controls* The controls must be set to the proper settings. If the controls are not set correctly, the appliance might not complete its cycle.

5. *Operate the appliance* Operate the appliance, and let it run through its cycle. Check the cycle operation against the operational sequence of events that is listed in the use and care manual.

6. *The appliance is operating properly* If it is, then explain to the consumer how to operate the appliance according to the manufacturer's specifications.

7. *The appliance is not operating properly* If the appliance is not operating properly, then proceed to locate which component has failed. Check the diagnostic charts that are listed in this book to assist you in the correct direction to take.

DIAGNOSIS AND CORRECTION PROCEDURE

When diagnosing a problem with the appliance, use your five senses to determine the condition of the appliance. This will help in analyzing and defining the problem.

- *Example #1* When turning on the washing machine, there is a smell of something burning. You can track down the location of the burning smell, and therefore, discover which part has failed.
- *Example #2* When turning on a dishwasher, unusual noises are heard coming from underneath the machine. Stop the dishwasher, and attempt to track down from where the noises are coming.

Along with your hand tools, there are a variety of test meters to assist in analyzing and defining the problem. Listed below are the sequence of events that should be taken when servicing an appliance.

- *Unplug the appliance* Change the range setting on the multimeter to voltage. Check the voltage from the appliance's receptacle. If there is an uncertainty, check the name plate rating, which is located on the appliance, for the correct voltage rating. When diagnosing a component failure there are three types of circuit failure. The open circuit, the grounded circuit, and the short circuit, all of which are thoroughly explained in the next chapter.
- *Gain access* Only remove the panels and screws necessary to gain access to where the suspected component failure is located.
- *Isolate and/or remove the defective part* Using the multimeter, isolate and/or remove the part, set the range to Ohms, and check for component failure. This will be further explained in chapter five.
- *Install the new component* When finding a defective part, replace it with a new original part. Reconnect all the wires in their original places.
- *Reattach all panels and screws* Now close the appliance, and reattach all panels and screws.
- *Test the appliance for proper operation* Plug in the appliance and test it.

5
Electricity

The technician must be knowledgeable in electrical theory to be able to diagnose and repair major appliances properly. Although this chapter cannot cover all there is to know about electricity, it will provide the basics. In the field of major appliances, the greatest number of potential problems are in the electrical portions of the appliance.

ELECTRICAL WIRING

The flow of electricity from a power source to the home can be made easier to understand by comparison to a road map. Electricity flows from a power source to a load. This is similar to a major highway, which runs from one location to another. High-voltage transformers are used to increase voltages for transmission over long distances. The power lines that go to different neighborhoods are like the smaller roads that turn off the major highway. The electricity then goes to a transformer that reduces the voltage going into the home. This is the intersection between the small roads and the medium-sized highways. The small road that goes into the neighborhood, and all the local streets, are like the wiring that goes inside the home.

When connecting all the streets, roads, and highways together, the city is accessible. That is similar to having electricity flowing from the power source to all of the outlets in the home.

Imagine driving down a road and coming to a drawbridge (in this case the switch) and it opens up. This stops the flow of traffic (electricity). In order for traffic (or electricity) to flow again, the drawbridge must close.

WHAT IS A CIRCUIT?

A *circuit* is a complete path through which electricity can flow, and then return to the power source. Figure 5-1 is an example of a complete circuit. To have a complete path (or *closed circuit*), the electricity must flow from point *A* to point *B* without interruption.

When there is a break in the circuit, the circuit is *open*. For example, a break in a circuit is when a switch is turned to its "off" position. This will interrupt the flow of electricity, or current, as in Fig. 5-2. When a broken circuit is suspected, it is necessary to discover the location of the opening.

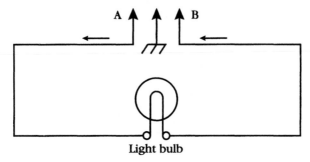

5-1 The complete circuit. Current flows from point A, through the light bulb, and back to point B.

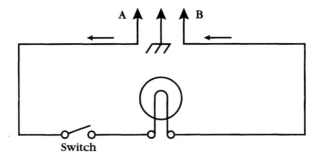

5-2 With the switch open, the current flow is interrupted.

CIRCUIT COMPONENTS

In an appliance, an electric circuit has four important components:

- *Power source* This source might be a battery, or the electricity coming from the wall outlet. Without the applied voltage, current cannot flow.
- *Conductors* A *conductor* will usually be a wire, and sometimes the metal chassis (frame). The function of the wire conductor is to connect a voltage source to a load.
- *Loads* These are the components that do the actual work in the appliance. A *load* is anything that uses up some of the electricity flowing through the circuit. For example, motors turn the belt, which turns the transmission. That, in turn, turns the agitator in a washing machine. Some other examples are heating elements and solenoids.
- *Controls* These control the flow of electricity to the loads. A control is a switch that is either manually operated by the user of the appliance, or operated by the appliance itself.

THREE KINDS OF CIRCUITS

You will come across three kinds of circuits:

1. Series circuits
2. Parallel circuits
3. Series-parallel circuits; a combination of series and parallel circuits.

Series circuit

In Fig. 5-3, the series circuit components are joined together in successive order, each with an end joined to the end of the next. There is only one path that electricity can follow. If a break is anywhere in the circuit, the electricity, or current flow, will be interrupted and the circuit will not function (Fig. 5-4). Figure 5-5 shows some of the many different shapes of series circuits, all used in wiring diagrams. In each series circuit, there is only one path that electricity can follow. There are no branches in these circuits where current can flow to take another path. Electricity only follows one path in a series circuit.

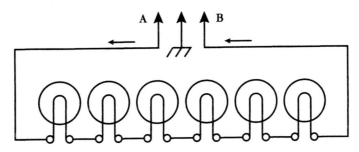

5-3 A series circuit.

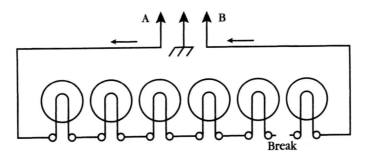

5-4 If there is a break in the wiring, all of the light bulbs are off.

Parallel circuits

In Fig. 5-6, the components are connected across one voltage source and form a parallel circuit. The voltage to each of these branches are the same. The current will also flow through all the branches at the same time. The amount of current that will flow through each branch is determined by the load, or resistance, in that branch. Figures 5-7 and 5-8 show examples of parallel circuits. If any branch has a break in it, the current flow will only be interrupted in that branch. The rest of the circuits will continue to function.

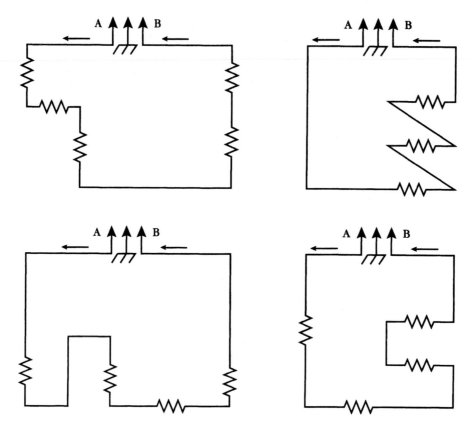

5-5 When you look at wiring diagrams, you will find series circuits in all sorts of shapes.

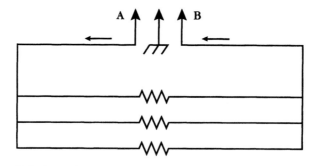

5-6 A parallel circuit.

Series-parallel circuits

A series-parallel circuit is a combination of series circuits and parallel circuits. In many circuits, some components are connected in series to have the same current, but others are in parallel for the same voltage (Fig. 5-9). This type of circuit is used where it is necessary to provide different amounts of current and voltage from the main source of electricity that is supplied to that appliance.

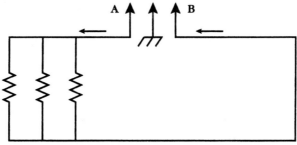

5-7 Another parallel circuit.

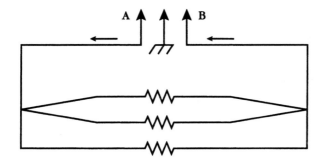

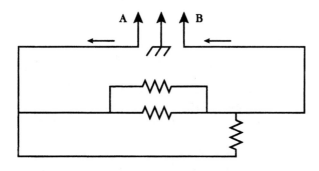

5-8
Notice that as in series circuits, the same parallel circuit can be drawn in many different ways.

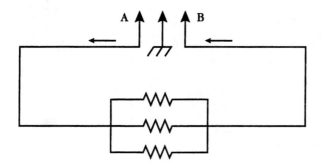

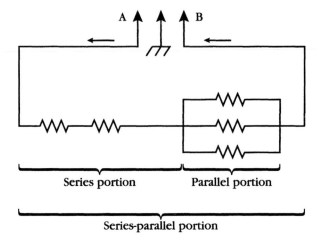

Series portion   Parallel portion

Series-parallel portion

5-9 Series-parallel circuit.

Series and parallel rules apply to this type of circuit. For example, if there is a break in the series portion of the circuit (Fig. 5-9), the current flow will be interrupted for the entire circuit. If the break is in the parallel portion of the circuit, the current will be interrupted for only that branch of the circuit. The rest of the circuits will still function.

TYPES OF ELECTRIC CURRENT

There are two types of electric current:
- *Direct current* Direct current, or dc, flows continuously in the same direction (Fig. 5-10).
- *Alternating current* Alternating current, or ac, flows in one direction, and then reverses itself to flow in the opposite direction, along the same wire. This change in direction occurs 60 times per second, which equals 60 Hz (Fig. 5-11).

Direct current (dc) is used in automobile lighting, flashlights, and cordless electric appliances; such as toothbrushes, shavers, drills, and in some major appliances.

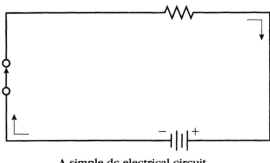

A simple dc electrical circuit

5-10
A simple dc electrical circuit. Current will flow from the negative side of the battery through the switch and load, and back to the positive side of the battery.

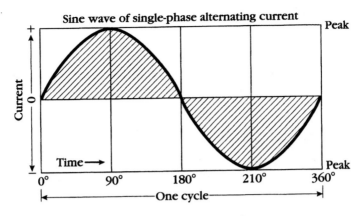

5-11 A waveform of a single-phase alternating current.

Alternating current (ac) is used in most homes. This current can be transmitted more economically over long distances than direct current. Alternating current can also be easily transformed to higher or lower voltages.

OHM'S LAW

Ohm's law states: The current which flows in a circuit is directly proportional to the applied voltage, and inversely proportional to the resistance. In other words: the greater the voltage, the greater the current; and, the greater the resistance, the less the current.

OHMS

Resistance is measured in ohms. Resistance opposes the flow of electrons (current). The amount of opposition to the flow is stated in ohms.

An instrument that measures resistance is known as an ohmmeter. Figure 5-12 is a schematic showing an ohmmeter connected to read the resistance of R1. The resistance of any material depends on the type, size, and temperature of its material. Even the best conductor offers some opposition to the flow of electrons. Figure 5-13 shows another type of meter, the multimeter, used for measuring ohms. The fundamental law to find resistance is stated: the resistance (R) in ohms is equal to the potential difference measured in volts (V) divided by the current in amperes (A). The equation is: $R = V \div A$.

AMPERES

Current is measured in amperes. The term "ampere" refers to the number of electrons passing a given point in one second. When the electrons are moving, there is current. Current can be measured in amperes, which is a measurement of the quantity of electron flow multiplied by time. The ammeter is calibrated in amperes, which we use to check for the amount of current in a circuit.

An instrument that will measure amperes is known as an ammeter. Figure 5-14 is a schematic showing an ammeter connected in a circuit to measure the current in amperes. Figure 5-15 shows an ammeter that is used in diagnosing appliance elec-

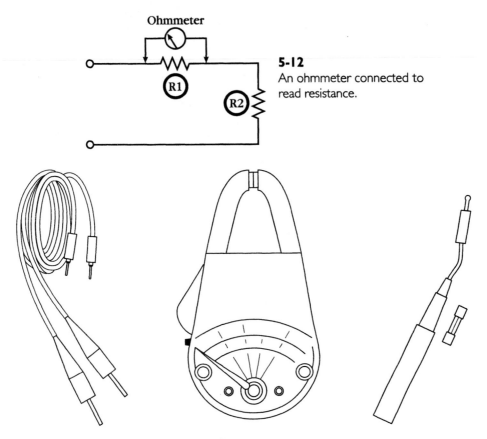

5-12
An ohmmeter connected to read resistance.

5-13 A test instrument for measuring ohms.

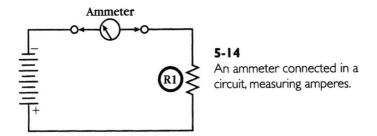

5-14
An ammeter connected in a circuit, measuring amperes.

trical problems. Current is the factor that does the work in the circuit (light the light, ring the buzzer). The fundamental law to find current is stated: the current in amperes is equal to the potential difference measured in volts divided by the resistance in ohms. The equation is: $A = V \div R$.

VOLTS

Electromotive force is measured in volts. This is the amount of potential difference between two points in a circuit. It is this difference of potential that forces current to

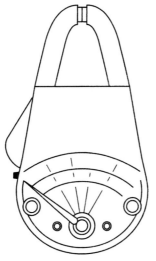

5-15
In the ammeter, the jaws clamp around a wire to measure the amperage of a circuit.

flow in a circuit. One volt (potential difference) is the electromotive force required to force one ampere of current through one ohm of resistance.

An instrument that will measure voltage is known as a voltmeter. Figure 5-16 is a schematic showing a voltmeter connected in the circuit to measure the voltage. Voltmeter 1 is connected to read the applied (or source) voltage. Voltmeter 2 is connected to measure the voltage drop, or potential difference, across R2. Figure 5-17 shows an actual volt-ohm-milliameter (VOM) that is used in measuring voltage.

The fundamental law to find voltage is stated: the potential difference measured in volts is equal to the current in amperes multiplied by the resistance in ohms. The equation is: $V = A \times R$.

WATTS

Power is measured in watts and an instrument that will measure watts is known as a wattmeter (Fig. 5-18). One watt of power equals the work done in one second, by one volt of potential difference, in moving one coulomb of charge. One coulomb per second is equal to one ampere. Therefore, the power in watts (W) equals the product of amperes times volts. The equation is: $W = A \times V$.

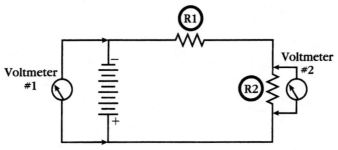

5-16 A voltmeter connected in a circuit to measure voltage.

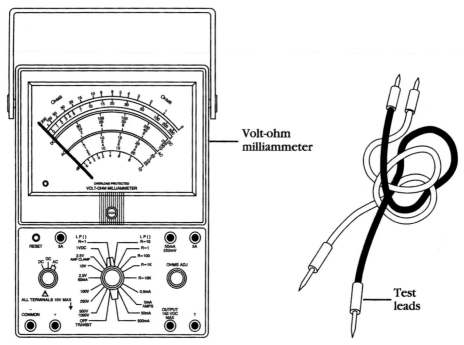

Volt-ohm
milliammeter

Test
leads

5-17 The volt-ohm-milliammeter with test leads.

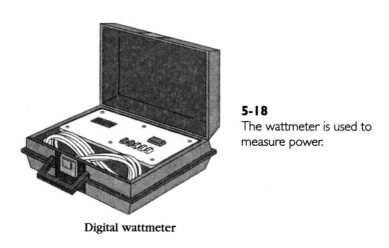

5-18
The wattmeter is used to
measure power.

Digital wattmeter

OHM'S LAW EQUATION WHEEL

The equation wheel in Fig. 5-19 shows the equations for calculating any one of the basic factors of electricity. Figure 5-20 shows the cross reference chart of formulas, as used in this text. If you know any two of the factors (V = voltage, A = amperage, R = resistance, W = power), you can calculate a third. To obtain any value in the center of the equation wheel, for direct or alternating current, perform the operation indicated in one segment of the adjacent outer circle.

Conversion chart for determining amperes, ohms, volts, or watts
(Amperes = A, Ohms = Ω, Volts = V, Watts = W

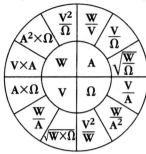

5-19 The Ohm's Law equation wheel.

Term	Measured in	Referred to in formulas as	Identification used in this text
Amperage	Amperes (Amps)	I	A for amps
Current	Amperes (Amps)	I	A for amps
Resistance	Ohms	Ω or R	R for resistance
Voltage	Volts	V or E	V for volts
Electromotive force	Volts	V or E	V for volts
Power	Watts	W	W for watts

5-20 The cross-reference chart of formulas.

Example: A 2400-watt heating element is connected to a 240-volt circuit. How many amps does it draw?

When finding amperage, the formula will be found in the amperes section of the wheel.

$$\frac{W\,(\text{watts})}{V\,(\text{volts})} = A\,(\text{amps})$$

Then, solving for amperage:

$$\frac{2400\,(\text{watts})}{240\,(\text{volts})} = 10\text{ amps}$$

What is the resistance (ohms)?

$$\frac{V^2\,(\text{volts squared})}{W\,(\text{watts})} = \text{ohms}$$

Then, solving for resistance:

$$\frac{240^2\,(\text{volts squared})}{2400\,(\text{watts})} = 24\text{ ohms}$$

WIRING DIAGRAM SYMBOLS

These wiring diagram symbols are commonly used in most wiring diagrams. Study each symbol so that you can identify them by sight (Tables 5-1, 5-2, 5-3, and 5-4).

Table 5-1

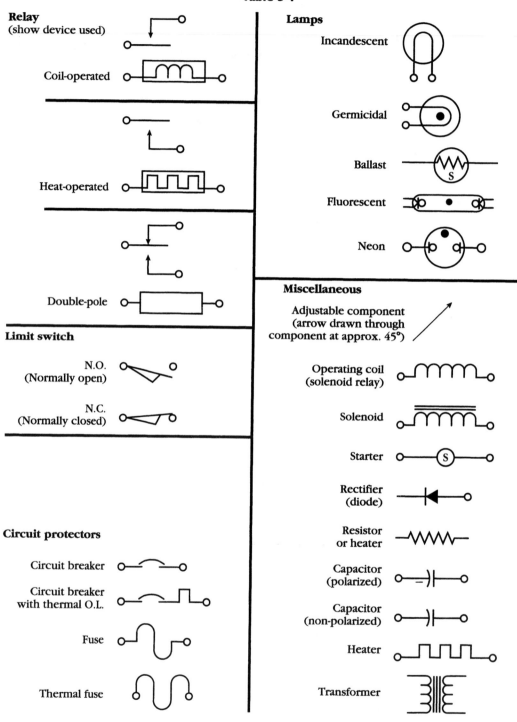

Relay
(show device used)

Coil-operated

Heat-operated

Double-pole

Limit switch

N.O.
(Normally open)

N.C.
(Normally closed)

Circuit protectors

Circuit breaker

Circuit breaker
with thermal O.L.

Fuse

Thermal fuse

Lamps

Incandescent

Germicidal

Ballast

Fluorescent

Neon

Miscellaneous

Adjustable component
(arrow drawn through
component at approx. 45°)

Operating coil
(solenoid relay)

Solenoid

Starter

Rectifier
(diode)

Resistor
or heater

Capacitor
(polarized)

Capacitor
(non-polarized)

Heater

Transformer

Table 5-2

Table 5-2

Temperature-actuated components

(Note: Symbols shown to be used for thermostats, bimetal switches, overload protectors, or other similar components, as required.

Temp. actuated
(close on heat rise)

Temp. actuated
(open on heat rise)

S.P.S.T.
(open on heat rise)

S.P.D.T.

S.P.D.T.

S.P.S.T.
(two contacts)

S.P.S.T. (adj.)
(close on heat rise)

S.P.D.T. (adj.)

S.P.S.T. (adj.)
(open on heat rise)

S.P.D.T. (adj.)
(with aux. "off" contacts)
(typical example)

S.P.S.T. (with internal heater)
(close on heat rise)

S.P.S.T. (with internal heater)
(open on heat rise)

Combination devices

Relay-magnetic
(arrangement of
contacts as necessary
to show operation)

Relay-thermal
(arrangement of
contacts as necessary
to show operation)

Timer (defrost)

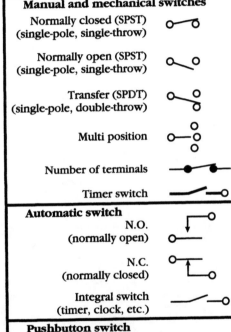

Manual and mechanical switches

Normally closed (SPST)
(single-pole, single-throw)

Normally open (SPST)
(single-pole, single-throw)

Transfer (SPDT)
(single-pole, double-throw)

Multi position

Number of terminals

Timer switch

Automatic switch

N.O.
(normally open)

N.C.
(normally closed)

Integral switch
(timer, clock, etc.)

Pushbutton switch
(momentary or spring return)

Circuit closing
N.O. (normally open)

Circuit opening
N.C. (normally closed)

Two circuit

SPDT
(single-pole, double-throw)

Table 5-3

Components		Lines and connections	
Buzzers		Integral conductor	
Adjustable		External or harness wire	
		Optional or alternate circuit	
Bell		Crossover ——+—— or	
		Permanent connection junction	
		Permanent connection	
Sensor (moisture)		Terminal	o
		Shield	
Thermocouple		Ground (earth)	
		Ground (chassis)	
Centrifugal switch		Grounded service cord (3-prong plug)	
Pressure switch S.P.D.T.		Service cord (2-prong)	
		Mechanical connection	
		Separable connector	

		Motors	
Humidistat		Timer or clock	TM
		Single-speed	
Magnetron		Two-speed	
		Three-speed	
Thermistor		Compressor	

Table 5-4

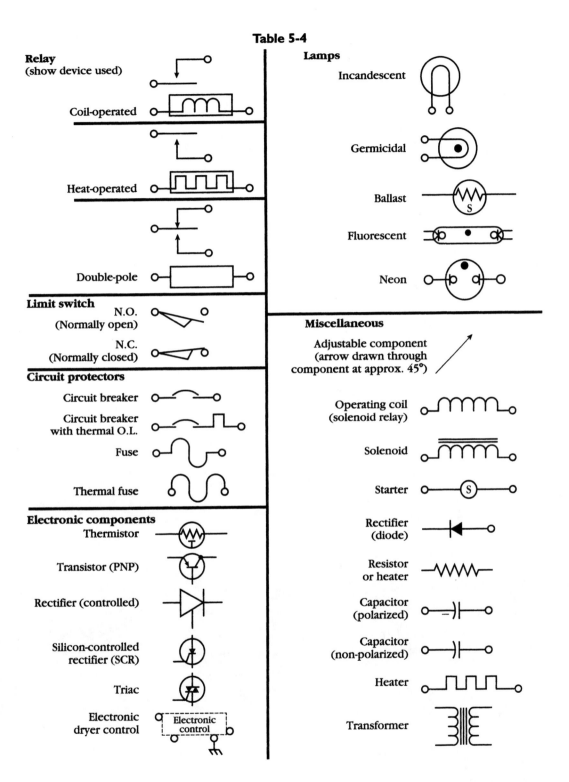

Relay
(show device used)

Coil-operated

Heat-operated

Double-pole

Limit switch

N.O. (Normally open)

N.C. (Normally closed)

Circuit protectors

Circuit breaker

Circuit breaker with thermal O.L.

Fuse

Thermal fuse

Electronic components

Thermistor

Transistor (PNP)

Rectifier (controlled)

Silicon-controlled rectifier (SCR)

Triac

Electronic dryer control

Electronic control

Lamps

Incandescent

Germicidal

Ballast

Fluorescent

Neon

Miscellaneous

Adjustable component (arrow drawn through component at approx. 45°)

Operating coil (solenoid relay)

Solenoid

Starter

Rectifier (diode)

Resistor or heater

Capacitor (polarized)

Capacitor (non-polarized)

Heater

Transformer

TERMINAL CODES

Terminal codes are found on all wiring diagrams. To help you identify the color codes they are listed in Table 5-5.

Table 5-5

Terminal color code	Harness wire color
BK	Black
BK-Y	Black with yellow tracer
BR	Brown
BR-O or BR-OR	Brown with orange tracer
BR-R	Brown with red tracer
BR-W	Brown with white tracer
BU or BL	Blue
BU-BK or BL-BK	Blue with black tracer
BU-G or BU-GN	Blue with green tracer
BU-O or BU-OR	Blue with orange tracer
BU-Y	Blue with yellow tracer
G or GN	Green
B-Y or GN-Y	Green with yellow tracer
G-BK	Green with black tracer
GY	Gray
GY-P or GY-PK	Gray with pink tracer
LBU	Light blue
O or OR	Orange
O-BK or OR-BK	Orange with black tracer
P or PUR	Purple
P-BK or PUR-BK	Purple with black tracer
P or PK	Pink
R	Red
R-BK	Red with black tracer
R-W	Red with white tracer
T or TN	Tan

Terminal color code	Harness wire color
T-R	Tan with red tracer
V	Violet
W	White
W-BK	White with black tracer
W-BL or W-BU	White with blue tracer
W-O or W-OR	White with orange tracer
W-R	White with red tracer
W-V	White with violet tracer
W-Y	White with yellow tracer
Y	Yellow
Y-BK	Yellow with black tracer
Y-G or Y-GN	Yellow with green tracer
Y-R	Yellow with red tracer

TIMER SEQUENCE CHARTS

Figure 5-21 represents a sample timer sequence chart. This detailed chart shows how the timer motor, and timer switch operation, control machine functions. When the timer switch sequence chart information is compared to the wiring diagram, electrical and mechanical diagnoses can be accomplished.

WIRING DIAGRAMS

In this segment, there are four different wiring diagram examples. Each example will take you through a step-by-step process in how to read wiring diagrams.

Example #1 Take a look at a simple wiring diagram for a refrigerator (Fig. 5-22). Note the black wire on the diagram. It is the wire that goes to the temperature control. The circuit is not energized when the temperature control is in the "off" position. When the temperature control knob is turned to the on position (switch contacts closed), and the circuit is energized; current will flow through the temperature control, through the red wire, through the overload protector, through the compressor and the relay, and back through the white wire to the line cord.

Example #2 Trace the active circuits with your finger. In the wiring diagram in Fig. 5-23, note switch number 1 on the diagram. It is the first switch in line, and it is the main switch that supplies voltage to the timer. No circuits are energized when the timer dial is in the off position (as the diagram indicates). When the user selects

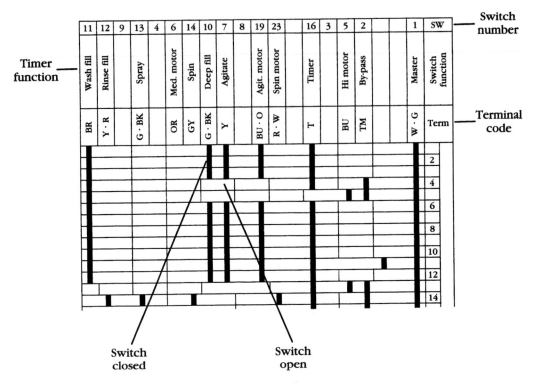

Switch number

Timer function

Terminal code

Switch closed / **Switch open**

5-21 A sample of a timer sequence chart.

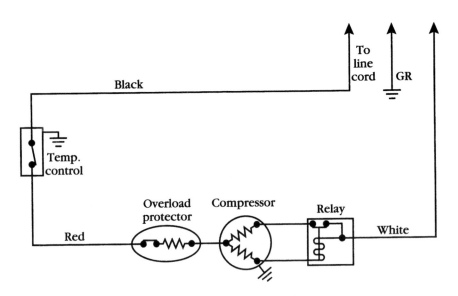

5-22 A simple wiring schematic of a refrigerator circuit.

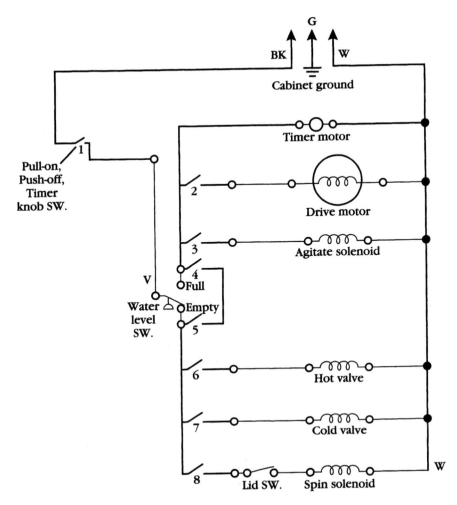

5-23 The wiring schematic of an automatic washer circuit.

the wash cycle with a warm wash, and turns on the timer, switch contacts 1, 2, 3, 6, and 7 are closed. (Use a pencil to close the switches on the diagram.) Voltage is supplied to the water level switch and the hot and cold water valves. Warm water is now entering into the washing machine tub. When the water level reaches the selected position, the water level switch contacts close from V to number 4, indicating that the water in the tub has reached the selected water level. When the water level switch is in this position the water is turned off. Voltage is now supplied to the timer motor, drive motor, and the agitate solenoid. The washing machine is now agitating and cleaning the clothes. As the timer advances to the spin cycle, timer switch contacts number 2 and 3 open; thus turning off the agitate solenoid and the drive motor. Switch contacts number 8 and 2 now close; thus supplying voltage to the drive motor and the spin solenoid. The timer motor advances to the end of the cycle. Timer switch contacts number 1, 2, and 8 will open. The washing machine is now off, no circuits are energized.

Example #3 The wiring diagram in Fig. 5-24 is for a refrigerator. Assume that the thermostat is calling for cooling, and the compressor in running. With your finger, trace the active circuits.

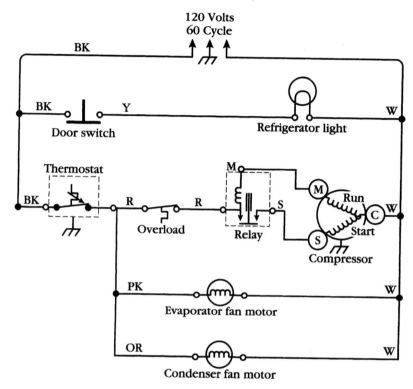

5-24 The wiring schematic of a refrigerator circuit.

The thermostat in the wiring diagram for the refrigerator is closed. The evaporator fan motor and the condenser fan motor are also running. Voltage is supplied through the overload to the relay. Current is flowing through the relay coil to the compressor-run winding. You will also notice that the door switch is open, and the refrigerator light is off. When the temperature in the refrigerator satisfies the thermostat, the thermostat switch contacts will open; thus turning off the compressor, the evaporator fan motor, and the condenser fan motor.

Example #4 The wiring diagram in Fig. 5-25 is for a no-frost refrigerator. Note the defrost timer in the lower left part of the diagram. The defrost timer switch contact is closed to contact number 4, the thermostat is calling for cooling, and the compressor is running. Trace the active circuits with your finger.

Voltage is supplied to the defrost timer terminal number 1. Current will flow through the defrost-2 timer motor to the white wire, and back to the line cord. At the same time, current flows through the number 4 contact in the defrost timer to the thermostat. At this point, the current passes through the thermostat to a junction, and splits in two directions. Current will flow through the temperature control,

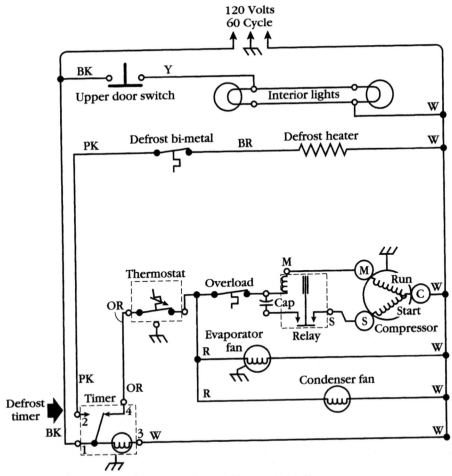

5-25 The wiring schematic of a no-frost refrigerator circuit.

through the overload protector, through the compressor and the relay, and back through the white wire to the line cord. At the same time, the compressor is running, current will flow through the evaporator, and condenser fan motors.

When the defrost timer activates the defrost cycle, the defrost timer switch contact is closed to contact number 2, and the compressor is not running. The evaporator and condenser fan motors will also stop running. The current will flow from terminal number 2, through the defrost bi-metal, through the defrost heater, and back through the white wire to the line cord. With your finger, trace the active circuits.

SAMPLE WIRING DIAGRAMS

The diagram in Fig. 5-26, is known as a ladder diagram. It is a simplified diagram using symbols, for parts and control components, attached to wires. The timer is represented by a dotted line running vertically through the switch contacts. To read and

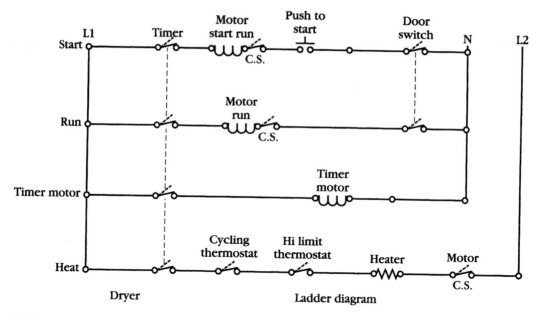

5-26 A simple ladder diagram.

understand this type of diagram, assume that the complaint is that the dryer is not heating. Look at the section of the diagram marked HEAT. Starting on the left side of the diagram, trace the circuit with your finger. You will notice that L1 goes to one side of the timer switch. As your finger moves from left to right, you will pass over the cycling thermostat, hi-limit thermostat, heater, motor centrifugal switch (CS), and on to L2. These are the components that make up the heating circuit. You must also include the motor and the door switch. If the door switch should fail, the motor will not run; thus opening the motor centrifugal switch (CS). If the motor fails, the high limit thermostat will open, shutting off the heater element. If any of these components fail, the dryer will not dry the clothing.

The diagrams in Figs. 5-27 and 5-28 illustrate a pictorial diagram and a schematic diagram. The pictorial diagram shows the actual picture of the components, and the schematic diagram uses symbols for the components.

Figure 5-29 shows a pictorial diagram of a refrigerator. In this type of diagram, you can see where the components are actually located.

HOW TO READ THE
VOLT-OHM-MILLIAMMETER TEST INSTRUMENT

The volt-ohm-milliammeter (VOM) is sometimes called a multimeter, because they can perform more than one function. A typical VOM will allow you to measure voltage, resistance, and current.

- Voltage equals electromotive force.
- Resistance equals the amount of resistance holding back the flow of current (measured in ohms).

- Current equals the amount of electricity flowing through a wire or circuit component (measured in amperes). There are many different types and brands of VOMs (Fig. 5-30). Most VOMs will have the following:
- Test leads. These are the wires coming from the meter to the part being tested.
- Meter scales and pointer (or a digital display on a digital meter). These show the amount of whatever value you are measuring.
- Function switch. This allows you to select whether you will be measuring ac or dc voltage (volts), current (amps), or resistance (ohms).
- Range selector switch. Allows you to select the range of values to be measured. On many meters, you can select functions and ranges with the one switch (as in the meter pictured in Figs. 5-30A and 5-30C). All VOMs are used in the same way to measure voltage, current, or resistance.

MEASURING VOLTAGE

If you don't have the right voltage in an appliance, the appliance won't function properly. You can find out whether the appliance is getting the right voltage by measuring the voltage at the wall outlet (receptacle). If an appliance isn't getting the proper voltage, nothing else you do to fix it will help.

For your safety, before using any test instrument, it is your responsibility as a technician to read and understand the manufacturer's instructions on how the test instrument operates.

Making the measurement

When measuring voltage, there are several steps you should follow:
- Attach the probes (another term for "test leads") to the meter. Plug the black probe into the meter jack marked negative or common. Plug the red probe into the positive outlet.
- Set the function switch to AC VOLTS.
- Select a range that will include the voltage you are about to measure. (Higher than 125 Vac if you are measuring 120 volts; higher than 230 Vac if you are measuring 220 volts.) If you don't know what voltage to expect, use the highest range, and then switch to a lower range if the voltage is within that lower range.
- Touch the tips of the probes to the terminals of the part to be measured.
- Read the scale.
- Decide what the reading means. Making voltage measurements is easy once you know how to select and read the scales on your meter.

SELECTING THE SCALES

To measure the voltage in appliances, use the ac/dc (Fig. 5-31). These same scales are used for both ac and dc readings.

The numbers on the right side of the ac/dc scales tell you what ranges are available to you. The meter face in Fig. 5-32 has three scales for measuring voltage. One is marked 10, another 50, and the third is marked 250. Remember, the scale you read is determined by the position of the range switch.

Example: If the range is set on 250 V, as in Fig. 5-33, you read the 0- to 250-V scale.

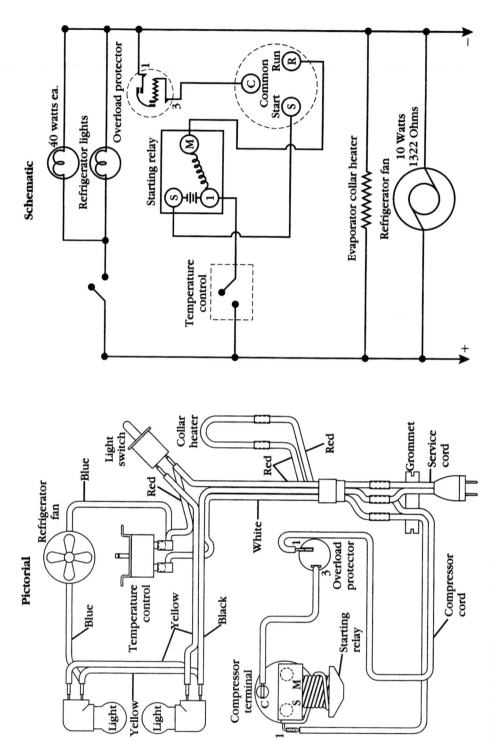

5-27 A pictorial diagram and a schematic diagram. Both diagrams show the same components.

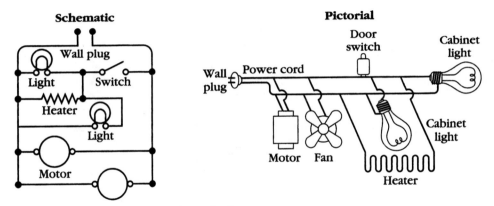

Schematic

Wall plug

Light   Switch

Heater

Light

Motor

Pictorial

Door switch

Cabinet light

Wall plug   Power cord

Cabinet light

Motor   Fan

Heater

5-28 A pictorial diagram and schematic diagram.

WIRE CODE	
COLOR	CODE
RED	RD
WHITE	WH
BLACK	BLK
ORANGE	OR
GREEN	GR
BLUE	BLU

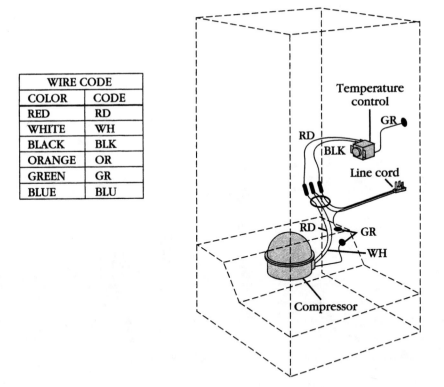

Temperature control

GR

RD

BLK

Line cord

RD   GR

WH

Compressor

5-29 A pictorial diagram of a refrigerator, showing where the components are located.

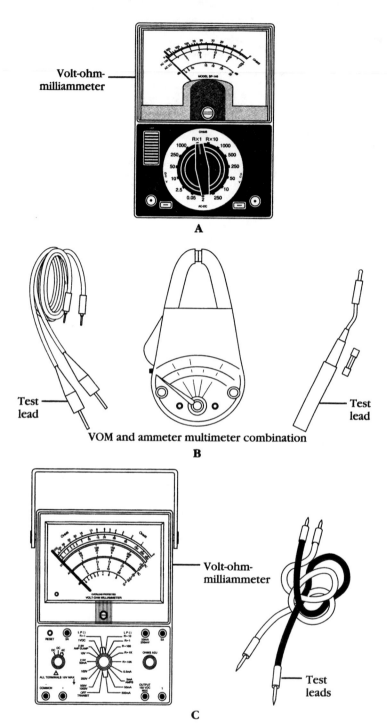

5-30 A. Volt-ohm-milliammeter. B. VOM and amperage multimeter combination. C. Volt-ohm-milliammeter.

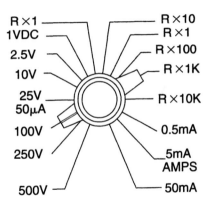

R ×1
1VDC
2.5V
10V
25V
50μA
100V
250V
500V

R ×10
R ×1
R ×100
R ×1K
R ×10K
0.5mA
5mA
AMPS
50mA

5-31
An example of scales used on some VOMs.

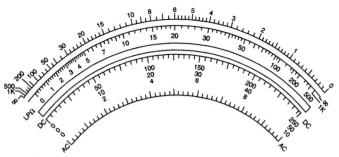

5-32 The meter face of an analog meter.

READING THE AC VOLTAGE SCALE

When reading the pointer position be sure to read the line marked AC (Fig. 5-32). The spaces on the voltage scales are always equally divided. When the pointer stops between the marks, just read the value of the nearest mark. In Fig. 5-34, the pointer is between 115 and 120 volts on the 250 scale. Read it as 120 volts. With an analog meter you're not gaining anything by trying to read the voltage exactly.

MEASURING LINE VOLTAGE

Measuring line voltage is the first, and most important, part of checking out an appliance that does not operate. Line voltage is the voltage coming from the wall outlet. There should be approximately 120 volts ac at the outlet under "no load" conditions. No load means that no appliance is connected, or that an appliance is connected, but it is turned off.

To measure line voltage—no load (120 volts) (Fig. 5-35):
- Set meter to measure ac volts.
- Set range selector to the range nearest to, but higher than, 120 volts.
- Insert either test lead into one slot of an empty wall receptacle.
- Insert the other test lead into the other slot of the same outlet. (Disregard the ground terminal for this test.) *Warning*!: Do not touch or handle the test leads by the metal portion of the probe. Hold the probe by the plastic grips that are attached to the test leads, to avoid electric shock.

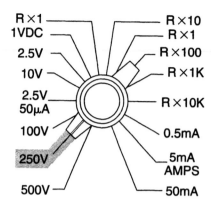

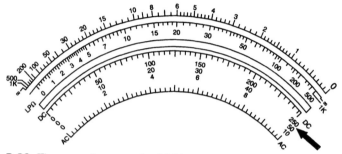

5-33 The range is set on the 250-V scale.

- Read the meter. The reading should be between 115 and 120 volts.
- When testing for 240 volts, be sure the range selector is set to the nearest range higher than 240 volts. Note: Most appliances are rated at 120 volts, but will work on voltages ranging from 110 to 125 volts. If the voltage drops more than 10%, the appliance will not operate; and most likely will damage some electrical components, if the appliance keeps running.

To measure line voltage—under load (120 volts) (Fig. 5-36):

- Be sure the appliance is plugged into one of the receptacles, and that the appliance is turned on.
- Follow steps 1 through 5 for no load conditions, inserting the test leads into the empty receptacle next to the one into which the appliance is plugged.
- Under load conditions (appliance is turned on), your reading will be slightly less than under no-load conditions.
- When testing for 240 volts, be sure the range selector is set to the nearest range higher than 240 volts. Note: Appliances with drive motors, such as automatic washers, dishwashers, and trash compactors, should also be tested at the moment of start. If the voltage drops more than 10% of the supplied voltage when the motor is started, it means that there is a problem with the electrical supply.

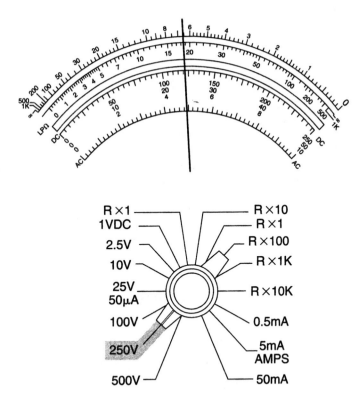

5-34 The range is set on the 250-V scale, the pointer indicates 120 V.

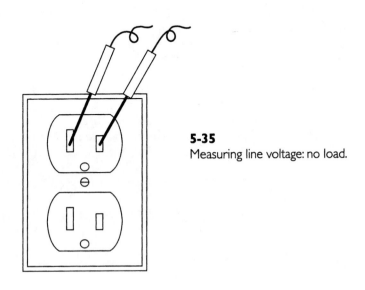

5-35
Measuring line voltage: no load.

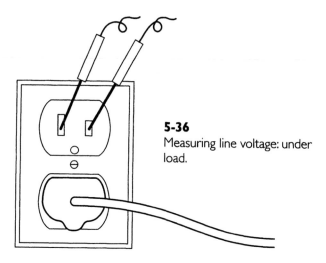

5-36
Measuring line voltage: under load.

TESTING FOR GROUND AND POLARITY

When a component is grounded to a chassis, there is no voltage between the component and the chassis. When a chassis is grounded to the earth, there is no voltage between the chassis and the earth.

If there is voltage between a chassis and the earth, it's dangerous. If you stand on the earth (dirt, concrete slab, etc.) and touch the chassis, electricity will flow through your body. You could injure yourself, or even someone else; and death could occur. It is important to be sure that the electrical outlets in the home, from which appliances are powered, must be correctly wired. That will protect the users from electrical shock.

If an outlet is wired "backward" (that is, if the black or red "hot wire" is connected to the long slot of the outlet), the appliance connected to that outlet might be unsafe to operate, or blow the fuse, or trip the circuit breaker. Appliances with solid state controls will not function properly if the outlets are wired backward.

To test for ground:
- First, test for line voltage. (If there is no voltage, you can't test for ground.)
- Notice that the receptacle has a longer and a shorter slot (Fig. 5-37). If the outlet has been mounted right-side-up, the longer slot will be on the left.
- Test for voltage between the short slot and the ground receptacle (the round hole) (Fig. 5-38). If there is line voltage between these two points, it means that the receptacle is grounded.
- If there is no round hole (ground prong receptacle), touch one of your probes to the screw that fastens the coverplate to the outlet (Fig. 5-39).

To test for polarity:
- Test to be sure that there is line voltage between the longer and shorter slots (Fig. 5-40).
- Test to be sure there is line voltage between the short slot and the center screw, or the round hole (ground prong receptacle) (Figs. 5-41 and 5-42).
- Test to be sure there is no voltage between the longer slot and the center screw, or the round hole (Figs. 5-43 and 5-44). Note: If these three tests don't test out this way, the outlet is incorrectly wired, and should be corrected by a licensed electrician.

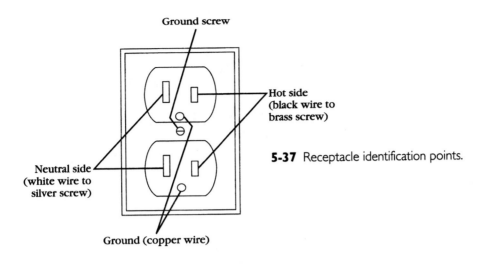

Ground screw

Hot side
(black wire to
brass screw)

Neutral side
(white wire to
silver screw)

Ground (copper wire)

5-37 Receptacle identification points.

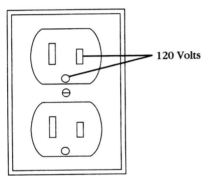

120 Volts

5-38 Testing for ground.

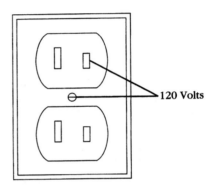

120 Volts

5-39 Testing for ground again.

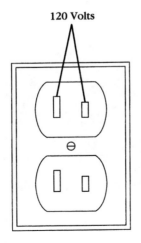

120 Volts

5-40
Testing for polarity.

Testing for ground and polarity 79

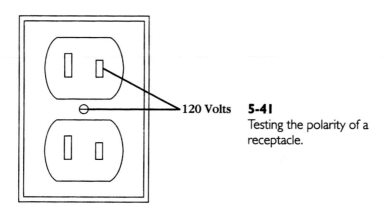

120 Volts

5-41
Testing the polarity of a
receptacle.

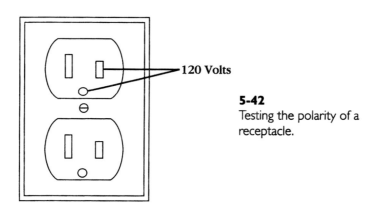

120 Volts

5-42
Testing the polarity of a
receptacle.

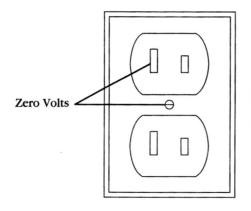

Zero Volts

5-43
Testing the polarity of a
receptacle.

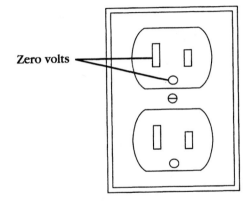

Zero volts

5-44
Testing the polarity of a receptacle.

MEASURING RESISTANCE (OHMS)

Electrical appliances need a complete path around which electricity can flow. If there is infinite resistance to the flow of electricity, you have an open circuit, or infinite resistance between the two points being measured.

When there is a complete path, you have continuity in the circuit. When you test to find out whether there is a break in the path, you say you are making a continuity check.

Continuity checks are made by measuring the amount of resistance there is to the flow of electricity. If there is so much resistance that it is too high to measure (called "infinite"), then you say that the circuit is open (there is no complete path for the electricity to follow). If there is some resistance, it means that there is continuity; but that there is also one (or more) load on the line—a light, a motor, etc. Note: A load is an electrical component that uses electricity to work (e.g., a light bulb, a motor, a heater coil).

If there is no resistance between the two points, it means the electricity is flowing directly from one point to the other. If the electricity flowed directly from one point to the other by accident or error, then you say you have a short circuit, or a "short."

SETTING UP THE METER

Measuring resistance (ohms) is like measuring voltage, except that the measurements are made with the electricity turned off. Listed below are the steps for measuring ohms:

- Attach the leads to the meter. Plug the black lead into the negative outlet, and the red lead into the positive.
- If your meter has a function switch, set the function switch to OHMS.
- If your meter has a range switch, set the range. The range selector switch will have several ranges of resistance (ohms) measurements (Fig. 5-45). The ranges are shown like this: R × 1: The actual resistance shown on the meter face, times 1. R × 10: The resistance reading, times ten; add one zero to the reading. R × 100: The resistance reading, times 100; add two zeros to the reading. R × 1K: The resistance reading, times 1000; add three zeros to the reading (K means 1000 Ω). R × 10K: The resistance reading, times 10,000; add four zeros to the reading.

- Set the range so it is higher than the resistance you expect. If you don't know what measurement to expect, use the highest setting, and adjust downward to a reading of less than 50. The left side of the scale is too crowded for an accurate reading.
- Zero the meter. You should "zero the meter" each time you set the meter. To "zero the meter" means to adjust the pointer so that it reads 0 Ω when the two test leads are touched together. Use the Ohms Adjust knob on the front of the VOM to line up the pointer over the zero on the ohms scale.
- Attach the test leads to the component you are measuring.
- Take the measurement. In Fig. 5-45, the range selector switch is on R × 100, and the measurement is 400 ohms.

ELECTRICAL SAFETY PRECAUTIONS

Know where, and how, to turn off the electricity to your appliance; for example: plugs, fuses, circuit breakers, or cartridge fuses. Know their location in the home. Label them. When replacing parts or reassembling the appliance, you should always install the wires on their proper terminals, according to the wiring diagram. Then, check to be sure the wires are not crossing any sharp areas, or pinched in some way, or between panels or moving parts that might cause an electrical problem. These additional safety tips can also help you and your family:
- Always use a separate, grounded electrical circuit for each major appliance.
- Never use an extension cord for major appliances.
- Be sure that the electricity is off before working on the appliance.
- Never remove the ground wire of a three-prong power cord, or any other ground wires, from the appliance.
- Never bypass, or alter, any appliance switch, component, or feature.
- Replace any damaged, pinched, or frayed wiring that might be discovered when repairing the appliance.

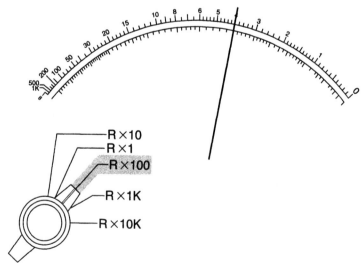

5-45 With the range set at R × 100, the meter reads 400x.

6

Appliance parts

Servicing a highly complex, electromechanical appliance is not as hard as might be expected. Just keep in mind that an appliance is simply a collection of parts, inside a cabinet, coordinated to perform a specific function (Figs. 6-1A, 6-1B, 6-1C, and 6-1D). Before servicing an appliance, you must know what these parts are, their functions, and their purposes.

THE SWITCH

The switch is a mechanical device used for directing and controlling the flow of current in a circuit. Simply put, the switch can be used for turning a component on or off (Fig. 6-2).

Internally, the switch has a set of contacts that close, allowing the current to pass; and when opened, forbidding the current to flow through it. Built into the switch, a linkage mechanism actuates these contacts inside of the closed housing

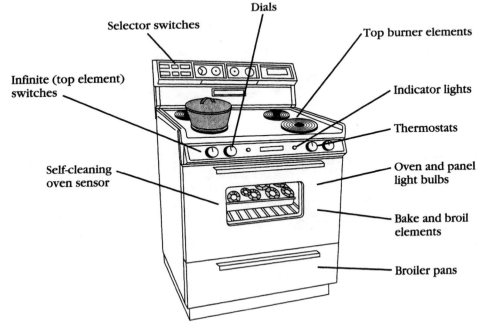

6-1A Typical electric range parts identification.

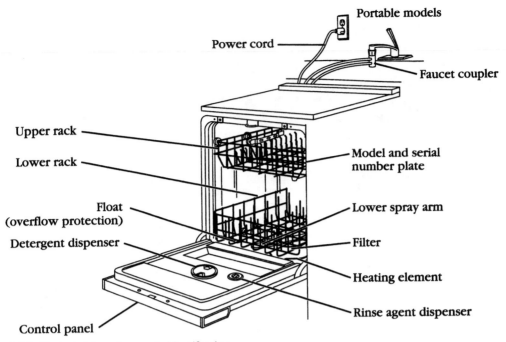

Portable models

Power cord

Faucet coupler

Upper rack

Lower rack

Float
(overflow protection)

Detergent dispenser

Control panel

Model and serial
number plate

Lower spray arm

Filter

Heating element

Rinse agent dispenser

6-1B Typical dishwasher parts identification.

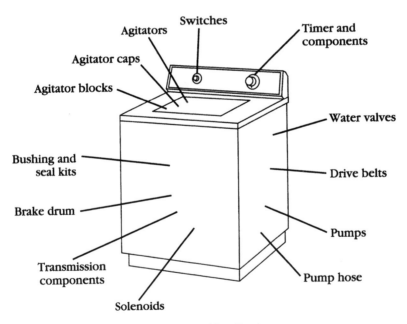

Switches

Agitators

Timer and
components

Agitator caps

Agitator blocks

Water valves

Bushing and
seal kits

Drive belts

Brake drum

Pumps

Transmission
components

Pump hose

Solenoids

6-1C Typical automatic washer parts identification.

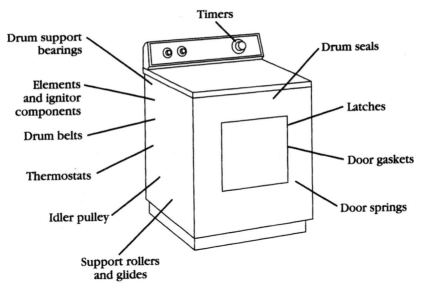

6-1D Typical automatic dryer parts identification.

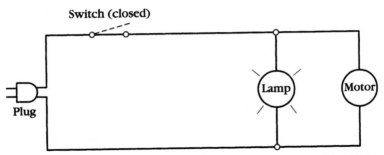

6-2 The wiring diagram illustrates a switch in the closed position. If the switch is closed, the light and motor are on. If the switch is open, the light and motor are off.

switch, a linkage mechanism actuates these contacts inside of the closed housing (Fig. 6-3).

Switches come in a wide range of sizes and shapes, and can be used in many different types of applications (Fig. 6-4). They have a voltage and amperage rating marked on the switch, or on the mounting bracket, for the type of service the individual switch was designed to do. The switch housing is usually marked with the terminal identification numbers that correspond to the wiring diagram. It identifies the contacts by number, normally open (NO) contacts, normally closed (NC) contacts, or common (COM). Internally, the switch can house many contact points for controlling more than one circuit.

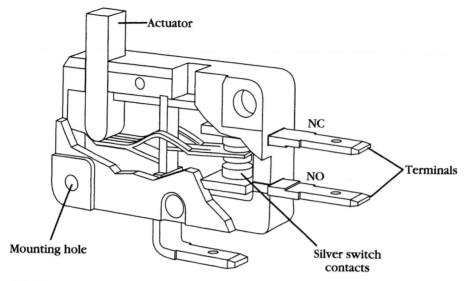

Actuator

NC

NO

Terminals

Mounting hole

Silver switch
contacts

6-3 The exploded view of a switch.

When suspecting a switch failure, remember that there are only three problems that can happen to a switch:

- The contacts of the switch might not make contact. This is known as an *open switch*.
- The switch's contacts might not open, causing a *shorted switch*.
- The mechanism that actuates the contacts might fail. This is a *defective switch*.

When these problems arise, the switches are not repairable, and they should be replaced with a duplicate of the original.

Pressure switch

The *pressure switch* is a specialty switch, with a similar operation to those above, but with one important exception. The pressure switch is actuated by a diaphragm that is responsive to pressure changes (Fig. 6-5). This switch can be found in washing machines and in dishwashers, and it operates as a water level control. Other uses include: furnaces, gas heaters, computers, vending machines, sump pumps, and other low pressure applications. The pressure switch is not serviceable, and should be replaced with a duplicate of the original.

Thermostat

The *thermostat* operates a switch. It is actuated by a change in temperature. The two most common heat-sensing methods, used in appliances, are the bimetal and the expansion types (Fig. 6-6).

The *bimetal type* (Fig. 6-7) consists of two dissimilar metals combined together as one. Any change in temperature will cause it to deflect, actuating the switch contacts. When the bimetal cools, the reverse action takes place.

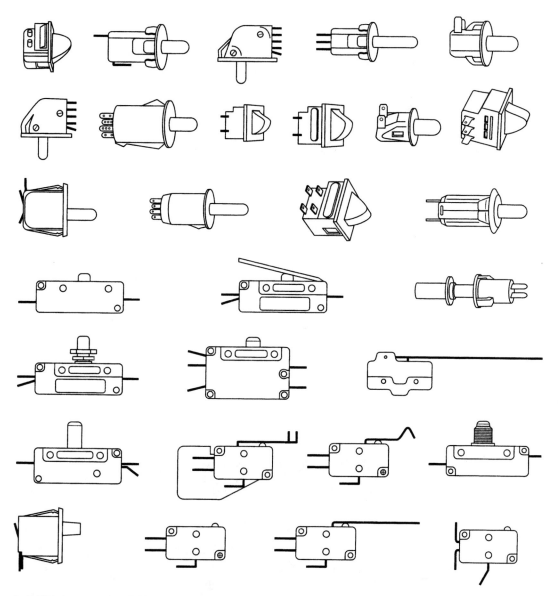

6-4 This is a sample of the many types of switches used in major appliances.

The *expansion type* (Fig. 6-8) uses a liquid in a tube that is attached to bellows. The liquid will convert to a gas when it is heated, and will travel up the tube to the bellows. This causes the bellows to expand, thus actuating the switch contacts. When the gas cools, the reverse action occurs.

Thermostats are used in applications as diverse as gas and electric ranges, automatic dryers, irons, waterbeds, spas, and in heating and refrigeration units.

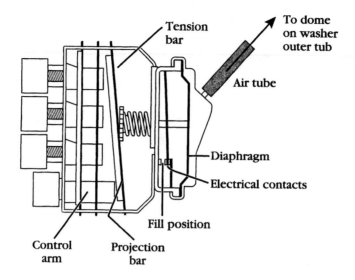

6-5 The pressure switch construction.

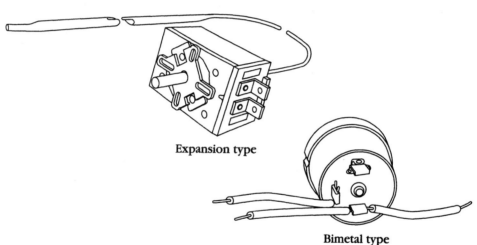

6-6 A bimetal and expansion-type thermostat.

Electromechanical timer

Although the timer is the most complex component in the appliance, don't assume that is the malfunctioning part. Check all of the other components associated with the symptoms, as described by the customer.

Electromechanical timers are utilized for controlling performance in automatic washers, automatic dryers, and dishwashers. Most timers are not serviceable, and should be replaced with a duplicate of the original.

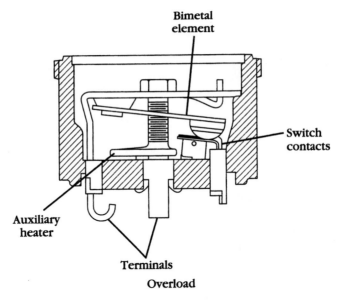

Bimetal element

Switch contacts

Auxiliary heater

Terminals

Overload

6-7 The bimetal thermostat construction.

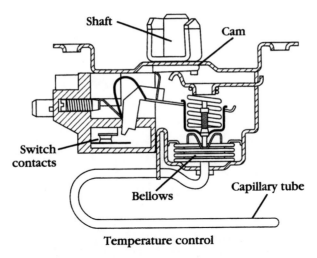

Shaft

Cam

Switch contacts

Bellows

Capillary tube

Temperature control

6-8 Temperature-control construction.

The timer assembly is driven by a synchronous motor, in incremental advances. It controls and sequences the numerous steps and functions involved in each cycle of an appliance (Fig. 6-9). The timer directs the on and off times of the components in an electrical circuit. The timer consists of three components assembled into one unit. The components are the motor, the escapement, and the cam switches (Fig. 6-10).

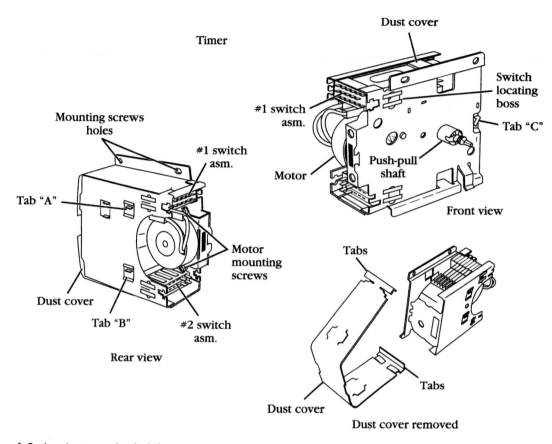

Timer

Dust cover

#1 switch asm.

Switch locating boss

Tab "C"

Push-pull shaft

Motor

Front view

Mounting screws holes

#1 switch asm.

Tab "A"

Motor mounting screws

Dust cover

Tab "B"

#2 switch asm.

Rear view

Tabs

Dust cover

Tabs

Dust cover removed

6-9 An electromechanical timer.

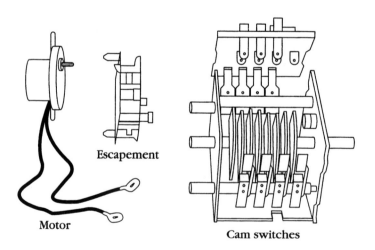

Escapement

Motor

Cam switches

6-10 The timer components: motor, escapement, and cam switches.

The motor

The motor is a *synchronous motor*, geared to drive the escapement. In Figs. 6-9 and 6-10, the motor is mounted on the timer assembly. These are specially designed motors, whose speed is controlled by the 60-cycle period of the current, rather than the fluctuating supply voltage.

The cam switches

The number of switch contacts, and cams, varies with the number of functions an appliance performs. Some of the switches perform two or more functions (Fig. 6-11). The particular shape of the cam varies with the number of switch contacts that it controls, and the length of time each switch contact opens and closes (Fig. 6-12). A single metal strip, called a cam follower (Fig. 6-11), is anchored to each cam switch arm. As the cam turns, this metal strip follows the contour of the circumference of the cam, causing the cam switch to open, or close, at the proper time.

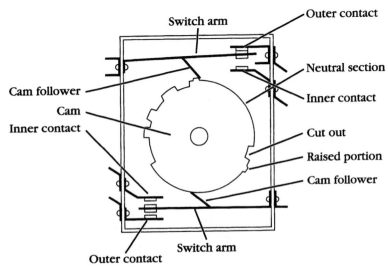

6-11 The timer cam and switch contacts' construction.

The escapement

The escapement is a spring-controlled mechanism that limits the cam shaft rotation to a set number of degrees in each increment (Fig. 6-10). The cam follower moves rapidly to ensure a snap action of the switch contacts, and also to prevent any arcing of the points. Not all timers have an escapement. For example, some dryer timers, and all defrost and electronic timers, have no escapement.

Relays

A relay is an electrically operated switch. It contains an electromagnet, with a fixed coil and a movable armature, that actuates a set of contacts to open and close electric circuits (Fig. 6-13). Relays have heavy-duty switching contacts, and they are used to operate larger components, such as motors and compressors. The most common failures of relays are:

- Open relay coil
- Burned switch contacts
- Armature not actuating the contacts. A broken relay is not serviceable, and should be replaced with a duplicate of the original.

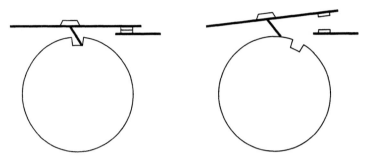

6-12 This illustration shows how a switch contact opens and closes on the cam.

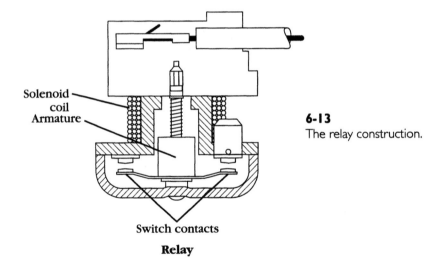

6-13
The relay construction.

Solenoid coil
Armature

Switch contacts

Relay

SOLENOIDS

A solenoid is a device used to convert electrical energy into mechanical energy (Fig. 6-14). When a solenoid is energized, it acts like an electromagnet, positioned to move a predesignated metal object. The work done by the moving core makes the solenoid useful in appliances. Some solenoids are equipped with a free-moving armature or plunger. Common failures of a solenoid are:
- Open coil.
- Shorted coil.
- Jammed armature.

A solenoid is not a serviceable part, and should be replaced with a duplicate of the original. These devices are manufactured in a variety of designs for various load

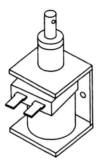

6-14
The solenoid coil and plunger.
When the coil is activated, the
plunger will be drawn to the
center of the magnetic field.

force and operational requirements. Solenoids are found in automatic washers and dryers, gas and electric ranges, automatic dishwashers, refrigerators, freezers, automatic ice machines, and in heating and air conditioning units.

WATER VALVES

The water inlet valve controls the flow of water into an appliance, and is solenoid-operated (Fig. 6-15). When it is energized, water in the supply line will pass through the valve body and into the appliance. Listed below are some different types of water inlet valves that are used on appliances:

- Single water inlet valve (Fig. 6-16); used on dishwashers, ice makers, refrigerators, undercounter ice machines.
- Dual water inlet valve (Fig. 6-17); used on washing machines, refrigerators, icemakers. Some dishwasher models also use dual water inlet valves. The inlet side of the valve has a fine mesh screen to prevent foreign matter from entering the valve. Some water valves also have a "water hammer" suppression feature built into them.

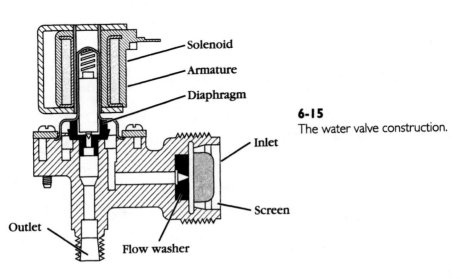

Solenoid
Armature
Diaphragm
Inlet
Screen
Outlet
Flow washer

6-15
The water valve construction.

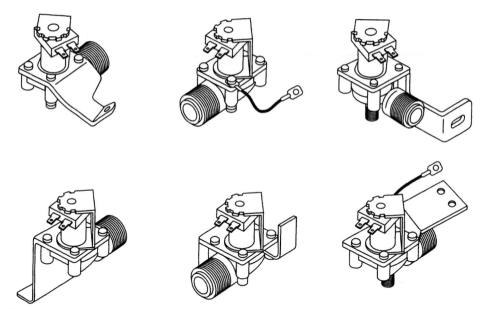

6-16 The single water valves are just some of the different types available that are used in major appliances.

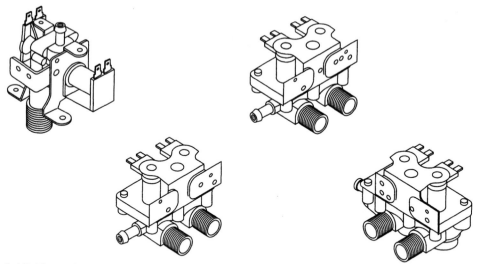

6-17 These dual water valves are designed for dual water-inlet connections.

Drain valves (Fig. 6-18) are used on some dishwasher and washing machine models to control the drainage of the water in the tub, and its expulsion into the sewage system of the residence.

The water valve should not be serviced. Replace it with a duplicate of the original.

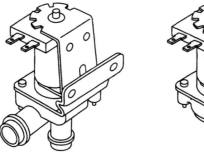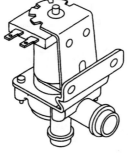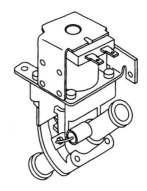

6-18 The drain water valves.

MOTORS

The two major assemblies that form an electric motor are the rotor and the stator (Fig. 6-19). The rotor is made up of the shaft, rotor core, and (usually) a fan. The stator is formed from steel laminations, stacked and fastened together so that the notches form a continuous lengthwise slot on the inside diameter. Insulation is placed so as to line the slots; and then coils, wound with many turns of wire, are inserted into the slots to form a circuit. The wound stator laminations are pressed into, or otherwise assembled within, a cylindrical steel frame to form the stator (Fig. 6-20). The end bells, or covers, are then placed on each end of the motor. One important function of the end bells is to center the rotor, or armature, accurately within the stator to maintain a constant air gap between the stationary and moving cores (Figs. 6-19 and 6-21).

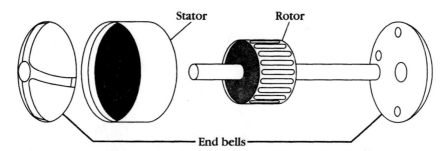

6-19 The stator and rotor.

These coils of wire are wound in a variety of designs, depending upon the electrical makeup of the motor. They provide two or more paths for current to flow through the stator windings. When the coils have two centers, they form a two-pole motor; when they have four centers, they form a four-pole motor. In short, the number of coil centers determines the number of poles that a motor has (Fig. 6-22).

Thermal protection in a motor is provided by a temperature-sensitive element, which activates a switch. This switch will stop the motor if the motor reaches the pre-set temperature limit. The thermal protector in a motor is a non-replaceable part,

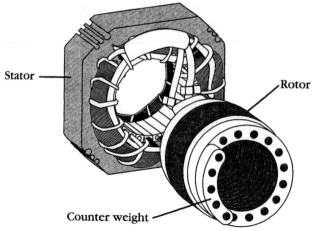

6-20 Stator and rotor construction.

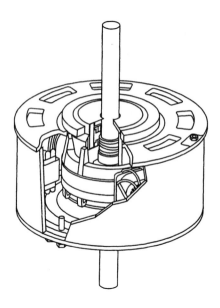

6-21 End bells position motor shaft in center of stator.

and the motor will have to be replaced as a complete component. There are two types of thermal protection switches:

- *Automatic reset* It automatically resets the switch, when the temperature has been reduced.
- *Manual reset* It has a small reset pushbutton on the motor, on the opposite end from the shaft.

On the following list are several types of motors that are used for different types of applications:

- Synchronous motors are permanent magnet timing motors, used in automatic ice cube makers, water softeners, and humidifiers. Also, they are integral to timers for automatic washers, automatic dryers, and dishwashers.

- Shaded pole motors are used as continuous duty motors, with limited or adjustable speeds. They are used for small fans and clocks.
- Split phase motors are used as continuous duty motors, with fixed speeds. Some are: automatic washer and dryer drive motors.
- Capacitor start motors are similar to the split phase motors, and they are used in hard-to-start applications such as compressors and pumps.
- Permanent split capacitor motors are used in a variety of direct-drive air-moving applications; for example, air conditioner fans.
- Three phase motors are used in industrial, or large commercial, applications where three-phase power is available.
- Multi-speed, split phase motors are used in fans, automatic dryers, automatic washers, and many other appliances.

Figure 6-23 illustrates some of these motors. Appliance motors are not repairable, and they should be replaced with a duplicate of the original.

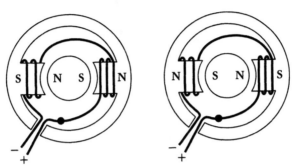

Two-pole stator motor.

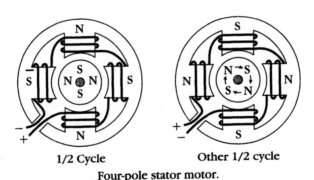

1/2 Cycle   Other 1/2 cycle

Four-pole stator motor.

6-22 2-pole motor/4-pole motor.

CAPACITORS

A capacitor is a device that stores electricity to provide an electrical boost for motor starting (Fig. 6-24). Most high-torque motors need a capacitor connected in series with the start winding to produce the desired rotation, under a heavy starting load. There are two types of capacitors:

- *Start capacitor* This type of capacitor is usually connected into the circuit between the start relay and the start winding terminal of the motor. Start capacitors are used for intermittent operation (on and off operation).
- *Run capacitor* The run capacitor is also in the start winding circuit, but it stays in operation while the motor is running (continuous operation). The purpose of the run capacitor is to improve motor efficiency during operation.

Capacitors are rated by voltage, and by their capacitance value in microfarads. This rating is stamped on the side of the capacitor. A capacitor must be accurately sized to the motor and the motor load. Both run and start capacitors can be tested by means of an ohmmeter.

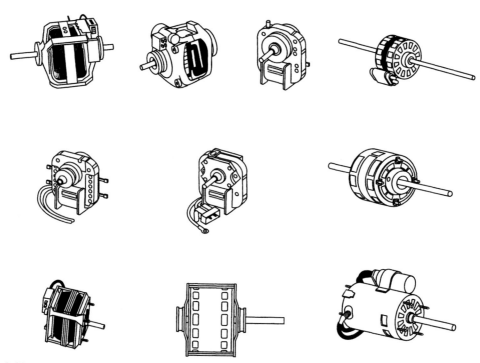

6-23 Motors are available in different sizes and shapes.

HEATING ELEMENTS

Most heating elements are made with a nickel-chromium wire having both tensile strength and high resistance to current flow. The resistance and voltage can be measured with a multimeter to verify if the element is functioning properly. Heating elements are available in many sizes and shapes (Fig. 6-25). They are used for:

- Cooking food
- Heating air for drying clothes
- Heating water to wash clothes, dishes, etc.
- Environmental heating.

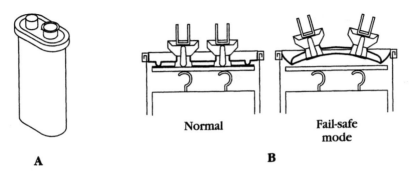

Normal   **Fail-safe mode**

A   **B**

6-24 A. The capacitor is rated by voltage and by capacitance (in microfarads). B. This built-in disconnect device is also known as a *fail-safe*.

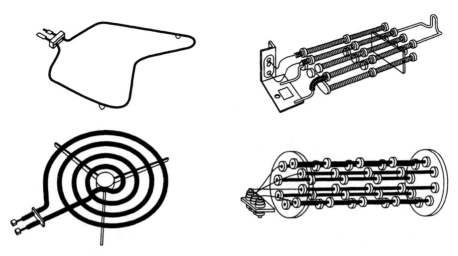

6-25 Heating elements.

Heating elements are not repairable, and they should be replaced with a duplicate of the original.

MECHANICAL LINKAGES

The mechanical linkages are those devices (connecting rods, gears, cams, belts, levers, pulleys, etc.) that are used on appliances in order to transfer mechanical energy from one point to another point. Figure 6-26, is an excellent example, the automatic ice maker. Some other examples are:

- In the automatic dryer, the motor is turning a pulley, which moves a belt, which turns the drum.
- In the automatic washing machine, the motor turns a pulley, which moves the belt, that turns the transmission gears, that performs the agitation or spin cycles.
- In the automatic icemaker, the timer gear turns the drive gear, which moves the cam, that actuates the switches, and rotates the ice ejector.

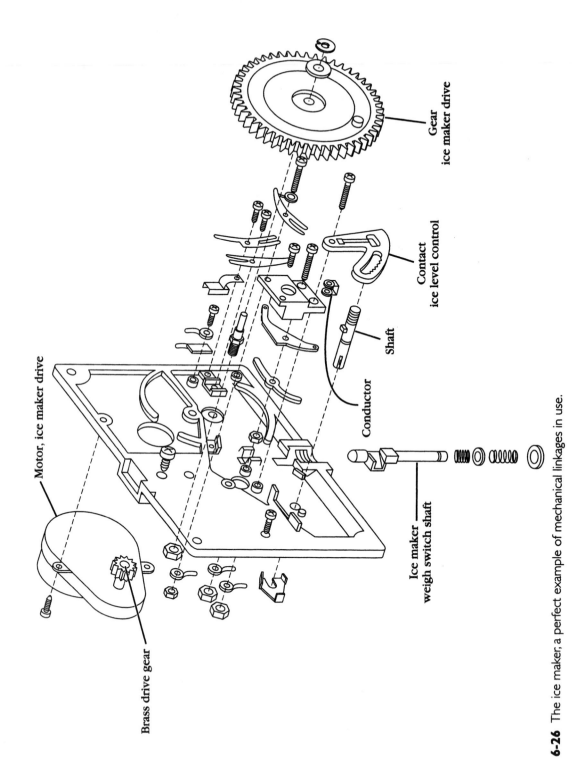

Gear
ice maker drive

Contact
ice level control

Shaft

Conductor

Motor, ice maker drive

Ice maker
weigh switch shaft

Brass drive gear

6-26 The ice maker, a perfect example of mechanical linkages in use.

WIRES

The wiring, which interconnects between the different components in an appliance, is the highway that allows current to flow from point A to point B. Copper and aluminum are the most common types of wires that are used in appliances. They are available as solid or stranded. Wires are enclosed in an insulating sleeve, which might be rubber, cotton, or one of the many plastics. Wires are joined together, or to the components, by:

- Solderless wire connectors
- Solderless wire terminal connectors
- Solderless multiple pin plug connectors
- Soldering

Never join copper and aluminum wires together, because the two dissimilar metals will corrode and interrupt the flow of current. The standard wire-gauge sizes for copper wire are listed in Table 6-1. As the gauge size increases from 1 to 20, the diameter decreases and the amperage capacity (ampacity) will decrease also (Table 6-2).

Table 6-1

Copper wire table

Gage no.	Diameter, mil.	Circular-mil area	Ohms per 1000 ft. of copper wire at 25° C*
1	289.3	83,690	0.1264
2	257.6	66,370	0.1593
3	229.4	52,640	0.2009
4	204.3	41,740	0.2533
5	181.9	33,100	0.3195
6	162.0	26,250	0.4028
7	144.3	20,820	0.5080
8	128.5	16,510	0.6405
9	114.4	13,090	0.8077
10	101.9	10,380	1.018
11	90.74	8234	1.284
12	80.81	6530	1.619
13	71.96	5178	2.042
14	64.08	4107	2.575
15	57.07	3257	3.247
16	50.82	2583	4.094
17	45.26	2048	5.163
18	40.30	1624	6.510
19	35.89	1288	8.210
20	31.96	1022	10.35

*20 to 25° C or 68 to 77° F is considered average room temperature.

Table 6-2

Size (AWG)	Ampacity
18	6
16	8
14	17
12	23
10	28

Part two

Appliance selection, service, installation, and preventive maintenance procedures

7
Automatic dishwashers

When a dishwasher is properly used, it will provide the customer with satisfactory results. There are times, however, when the dishwasher is blamed for poor performance. Perhaps the customer does not know how to load the dishwasher properly, or they used the wrong amount of detergent, and in some cases the dishwasher might not run at all. Whatever the case might be, it is up to the technician to repair the dishwasher, or to instruct the owner in its proper usage.

This chapter provides the technician with the basic skills needed to diagnose and repair automatic dishwashers.

PRINCIPLES OF OPERATION

After placing the dishes properly in the dishwasher, the detergent is placed in the dispenser, and the rinse conditioner is checked for the proper level. The door is closed, and the type of wash cycle is selected.

The door latch holds the door closed and activates the door latch switch. This will complete the electrical circuit for the dishwasher to operate. If the door is opened during the cycle, this will cause all operations to cease.

The timer will energize the water inlet valve. Water will begin to enter the tub. The dishwasher does not fill with water like a washing machine. It is designed so that the tub does not have more than two gallons of water in it at any one time. Should the timer switch contacts fail to open during the fill cycle, a float switch assembly, located inside the tub, will open the electrical circuit to the water inlet valve at a preset level.

The fill safety switch is part of the float assembly. Should the timer fail to open its switch contacts, water will keep entering the tub until the float, located inside the tub, rises and engages the float switch to shut off the water. Note: The float switch will not protect against a mechanical failure of the water inlet valve.

1. During the wash and rinse portion of a cycle, the heater element heats the water (on some models) to at least 140 degrees Fahrenheit. This feature is built into the dishwasher, and it is designed to save the consumer money in the operating costs of washing the dishes. Also, the customer does not have to raise the water temperature of the water heater to 140 degrees Fahrenheit. They only need to set the water heater temperature at 120 degrees Fahrenheit. This will prevent any member of the household from getting burned.

2. The water is pumped through the lower and upper spray arms, and onto the dishes, over and over again.

As the water runs off the dishes and back to the pump, it flows through a filtering system. On some models, the filter is designed to separate most food particles from the water so they won't be sprayed back onto the dishes.

At the end of a wash or rinse cycle, the water is pumped out of the dishwasher, flushing the filter of any small food particles. The larger pieces of food are trapped on the pump guard, which must be cleaned out before the next use.

On some models, the pump screen removes food particles from the water, stores them, and then grinds up the food particles as they are washed down the drain. During grinding, some sounds will be heard.

At the end of the cycle, the heater element comes on (if this is selected) and helps dry the dishes. Certain models have a fan that circulates the air to speed up the drying cycle, thus making sure that all dishes are evenly dry. Some models have a cool dry cycle. This cycle allows the dishes to be dried without the heater element operating. Combined with the heated air within the tub (from the wash cycle) and the dishwasher door opened a little bit, it will cause the water that remains on the dishes to condense and roll off the dishes. Figure 7-1 illustrates component locations within the automatic dishwasher.

SAFETY FIRST

Any person who cannot use basic tools, or follow written instructions, should *not* attempt to install, maintain or repair an automatic dishwasher. Any improper installation, preventive maintenance, or repairs could create a risk of personal injury or property damage.

If you do not fully understand the installation, the preventive maintenance, or the repair procedures in this chapter, or if you doubt your ability to successfully complete the task on the automatic dishwasher, then please call your service manager.

Before continuing, take a moment to refresh your memory of the safety procedures in Chapter 2.

DISHWASHERS IN GENERAL

Much of the troubleshooting information in this chapter covers the various types of dishwashers in general, rather than specific models in order to present a broad overview of service techniques. The pictures and illustrations that are used in this chapter are for demonstration purposes only; they clarify the description of how to service an appliance and they in no way reflect a particular brand's reliability.

Location and installation of a dishwasher

Locate the dishwasher where there is easy access to existing drain, water and electrical lines (Fig. 7-2). Be sure to observe all local codes and ordinances for electrical and plumbing connections. It is strongly recommended that all electrical and plumbing work should be done by qualified personnel. The best location for the dishwasher is on either side of the sink.

For proper operating and appearance of the dishwasher, the cabinet opening should be square, and have dimensions as shown in Fig. 7-3. If the dishwasher is

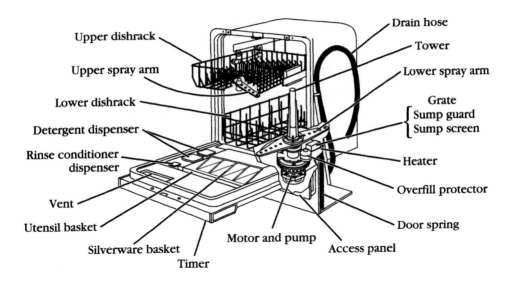

Upper dishrack
Upper spray arm
Lower dishrack
Detergent dispenser
Rinse conditioner dispenser
Vent
Utensil basket
Silverware basket
Timer
Motor and pump
Access panel
Door spring
Overfill protector
Heater
Grate { Sump guard Sump screen
Lower spray arm
Tower
Drain hose

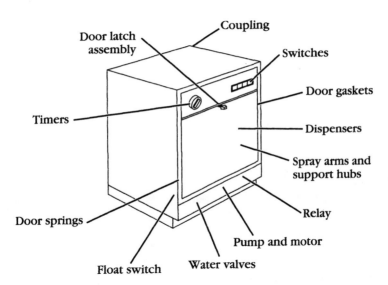

Door latch assembly
Coupling
Switches
Door gaskets
Dispensers
Spray arms and support hubs
Relay
Pump and motor
Water valves
Float switch
Door springs
Timers

7-1 Dishwasher component location.

to be installed in a corner, there must be sufficient clearance to open the door (Fig. 7-4).

Take the time to read over the installation instructions, and the use and care manual, that comes with every new dishwasher. These booklets will provide you with very important information. Such as:

- Safety
- Tools needed for the installation
- How to remove the panels
- How to change the color of the panels

- Locating drain, water, and electrical supply
- Position, align, and level the dishwasher
- Drain hose connection
- Water line connection
- Connecting the dishwasher to the electrical supply
- Securing the dishwasher in the opening you have selected
- Proper operation of the dishwasher
- Most importantly, read the warranty information that is supplied with the dishwasher

Functions and cycles

Dishwashers are very similar to automatic clothes washers. They apply three kinds of energy on the things to be washed. These forces are:
- *Mechanical energy* Water that is sprayed onto the dishes to remove the food particles.
- *Heat energy* Using hot water to liquefy the fats and greases on dirty dishes.
- *Chemical energy* Detergents to dissolve the fats and greases off the dishes.

Dishwashers perform four basic functions that are modified and put together in different ways to create the various cycles. The four functions are:
- Fill
- Wash/rinse
- Drain
- Dry

As with clothes washers, the only difference between the wash and rinse cycles is the presence of detergent in the wash water. The mechanical activities that make up a wash and a rinse cycle are basically the same function.

Unlike clothes washers, most dishwashers fill and begin to wash (or rinse) at the same time. The functions are put together in various ways to make up different cycles. By example, some of those cycles are:
- *Normal wash cycle* A single or double wash, with several rinses and a dry.
- *Heavy wash cycle* Adds a wash to the normal wash cycle.
- *Pots and pans* Similar to the heavy wash cycle, but this cycle heats the water in the wash cycles, and (on some models) this also heats the rinse cycles. On certain models, the timer will not advance until the water temperature is 140 degrees Fahrenheit. This will extend the total time of operation. Depending on the model, the cycle time can increase from 15 minutes to 2 hours. (Check the use and care manual)
- *Light wash cycle* This cycle is like the normal wash cycle, but minus a wash cycle.
- *Rinse and hold* Two rinses for holding dishes to wash later.

Water temperature

The temperature of the incoming water is critical to the operation of a dishwasher. Most dishwashers have heaters, and some have delay periods that extend the time periods during which water is heated to a specified point, but this does not fully compensate for low temperature of the water supply. You can check the tempera-

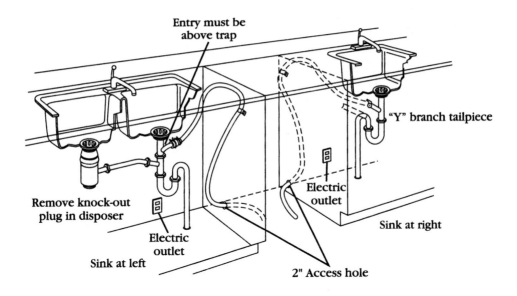

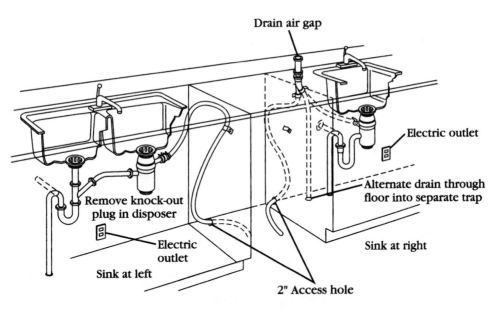

7-2 Typical dishwasher installation for a left or right sink application.

ture of hot water at the sink nearest to the dishwasher with a thermometer. Open the hot water faucet. Let the water run until it is as hot as possible, then insert the thermometer into the stream of water. On some models, if the thermometer reading is below 140 degrees Fahrenheit, then you will have to raise the water heater thermostat setting. On other models, the dishwasher was designed to operate with

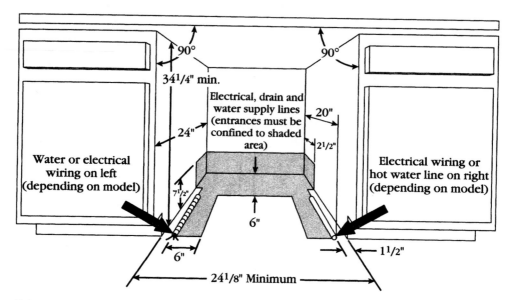

7-3 Undercounter dishwasher cut-out dimensions.

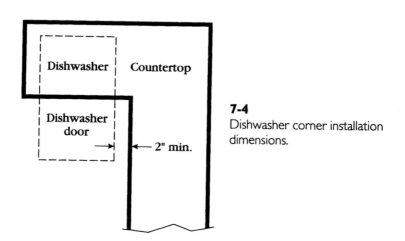

7-4
Dishwasher corner installation dimensions.

water temperatures as low as 120 degrees Fahrenheit. These models have longer detergent wash periods that allow 120-degree Fahrenheit supply water to be heated up to a temperature that gives good washability results. The dishwasher delay periods occur in only one, two or three of the water fills, and do little for the remaining rinses. Except during delay periods, the water is not in the dishwasher long enough to be heated adequately.

Water temperature above 150 degrees F
It is not recommended to have the water temperature above 150 degrees Fahrenheit in a domestic dishwasher; above this temperature, certain components in the dishwasher might be adversely affected.

Water temperature at 150 degrees F

Water temperature 150 degrees Fahrenheit is the ideal temperature for a mechanical dishwasher. Detergent action, and the dissolving of greases, are at the maximum at this temperature. Drying of most materials in the dishwasher will be very satisfactory.

Water temperature at 140 degrees F

Water temperature at 140 degrees Fahrenheit is the minimum temperature recommended by most dishwasher and detergent manufacturers. At this temperature, detergent is still quite active, and most fats are dissolved so that they can be emulsified in the water by the detergent, and washed down the drain. Drying will be fair to poor as water temperature in the last rinse is lowered toward 140 degrees Fahrenheit in some models. Some improvement in drying is possible if the user will add a liquid wetting agent to the dispenser.

Water temperature between 130 and 140 degrees F

Water temperature between 130 and 140 degrees Fahrenheit is outside of the range for best dishwasher operations, and users will have to exert special care if they are to obtain satisfactory results. The cleaning action of detergents, and the dissolving of fats, are gradually reduced as temperatures drop below 140 degrees Fahrenheit, so the dishes will probably have to be well rinsed before putting them into the dishwasher. Satisfactory drying becomes less likely as water temperature becomes lower. Some dishwashers have an optional feature which will increase the temperature of the water in the tub at different points during the cycle.

Water temperature between 120 and 130 degrees F

Water temperatures between 120 and 130 degrees Fahrenheit will aggravate all of the conditions and problems mentioned for the 130- to 140-degree range. Very few fats will dissolve, so the greasy buildup in the lower areas of the tub will be accelerated. Sudsing and foaming is more likely to increase. Detergent action is further reduced, so pre-rinsing of the dishes becomes even more important. The water heat feature will most likely have to be used.

Water temperature below 120 degrees F

Because of poor washing, grease buildup, poor drying and foaming, it is most unlikely that the dishwasher will perform to the satisfaction of the user, if the water supply is at less than 120 degrees Fahrenheit.

The detergent

The kind and amount of dishwasher detergent that is used is an important part of getting the dishes clean. Different brands of dishwasher detergent contain different amounts of phosphorous, which works to soften water and prevent water spots. If the water is hard, you will have to instruct the customer to use a detergent with a higher phosphorous content above 12%. If the water is soft, the customer can use a low-phosphorous dishwasher detergent. Some areas restrict the phosphate content to 8%, or even less. This means that the customer will have to increase the amount of detergent used, in those areas where the water is hard. This is done by adding 1 teaspoon of dishwasher detergent manually, in the main wash cycle, for each grain of water hardness above 12 grains. Water hardness is measured in grains.

- 0 to 3 grains for soft water
- 4 to 9 grains for medium hard water
- 10 to 15 grains for hard water
- over 15 grains for very hard water. If the hardness of the water supply is unknown, contact the local water department.

Always instruct the user to use automatic dishwasher detergent only. The use of soap, hand dishwashing detergent, or laundry detergent will produce excessive suds and will cause flooding and damage to the dishwasher.

STEP-BY-STEP TROUBLESHOOTING BY SYMPTOM DIAGNOSIS

In the course of servicing an appliance, you might overlook the simple things that might be causing the problem. Step-by-step troubleshooting, by symptom diagnosis, is based upon diagnosing the malfunctions, with possible causes arranged into categories relating to the operation of the dishwasher. This section is intended only to serve as a checklist to aid in diagnosing a problem. Look at the symptom that best describes the problem that you are experiencing with the dishwasher, then proceed to correct the problem.

Before testing any electrical component for continuity, disconnect the electrical supply to the appliance.

No water to dishwasher

1. Is the water turned on?
2. Is there voltage to the water inlet valve solenoid?
3. Is the water inlet valve solenoid defective? Disconnect the electrical supply, and check the solenoid coil with an ohmmeter.
4. Is the water valve plunger stuck? Disassemble the water valve and check.
5. Is the water valve inlet screen blocked? Turn off the water supply and remove the water supply line to inspect the screen.
6. Is the water fill line that goes to the water valve kinked? Check visually.
7. Is the water siphoning out of the dishwasher while it is filling? Check the drain hose installation.

Low water charge

1. Is there adequate water pressure to the water inlet valve? Pressure should be between 15 and 120 pounds per square inch.
2. Is the volume adequate? Take a quart container and fill it at the tap. The water must fill the container in seven seconds or less.
3. Is the water valve inlet screen clean? Turn off the water supply and remove the water supply line to inspect the screen.
4. On portables, you will check the aerating snap adapter on the faucet.
5. Is the timer defective? Disconnect the electric supply and check the switch contacts for continuity.
6. Is the float switch improperly positioned? Is the switch defective? Disconnect the electric supply and check the switch contacts for continuity.
7. Is the water fill line that goes to the water valve kinked? Check visually.

Poor washability on the upper rack

1. Is the upper spray arm turning?
   a. Are the holes in the spray arm plugged?
   b. Check to see if the spray arm is split.
   c. Is there uneven loading of the dishes?
   d. Check the lower impeller to see if it is defective, or blocked with debris.
   e. Are any objects protruding down which might prevent the upper spray arm from rotating?
2. Is the water charge OK?
3. What is the temperature of the water entering the tub? Is the temperature at 140 degrees Fahrenheit?
4. Is the consumer using the proper amount of detergent?
5. Is the detergent dispenser functioning properly?
6. Are the dishes loaded properly? Ask the user to load the dishwasher so that you can observe whether the user is loading the dishwasher properly.

Poor washability in the lower rack

1. Is the lower spray arm turning?
   a. Are the holes in the spray arm plugged?
   b. Check to see if the spray arm is split.
   c. Is there uneven loading of the dishes?
   d. Is the spray arm binding on the housing?
   e. Are any objects protruding down which might prevent the lower spray arm from rotating?
2. Is the water charge OK?
3. What is the temperature of the water entering the tub? Is the temperature at 140 degrees Fahrenheit?
4. Is the consumer using the proper amount of detergent?
5. Is the detergent dispenser functioning properly?
6. Are the dishes loaded properly? Ask the user to load the dishwasher so that you can observe whether the user is loading the dishwasher properly.

Poor drying

1. Is the water hot enough? Check the water supply temperature. On certain models, check to see if the cycle extender is functioning properly.
2. Is the heater working?
   a. Check the wiring
   b. Check the timer
   c. Check for the correct voltage
   d. Check the heater itself
   e. Check the heater fan assembly
3. Is a wetting agent being used?
4. Check the door baffle.
5. Suggest to the user to open the door a little after the dry cycle has been completed.

Water leaks at the front of the dishwasher

1. Is the spray arm turning? Is it split? Are any of the holes blocked with debris? If so, this can cause the water to spill out of the tub.
2. Is the tub gasket in place? Check the corners especially. Worn tub or corner gaskets will cause water to leak out of the corners.
3. Did you check the vent baffle gasket?
4. Check to see if the tub is not overcharged with water. Too much water will spill over the front tub flange.
5. Does the door close properly and tightly where the latch fits the strike?
6. On some models, are the gasket clips in place? These clips secure the gasket to the tub.
7. Is the dishwasher loaded properly? Ask the user to load the dishwasher so that you can observe whether the user is loading the dishwasher properly.
8. Check to see if the dishwasher is draining properly. Inspect the drain hose where it enters the drain.
9. Most importantly, is the dishwasher level?
10. Check the door for any cracks.
11. Check to see if the wetting agent dispenser is working properly. Too much wetting agent will cause sudsing. Also, check the gasket between the door and the dispenser.

Water leaks at sides, top, or bottom of dishwasher

1. Check the side inlet tube on the side of the dishwasher.
2. Check to see if water is leaking out from the pump assembly.
3. Check the motor seals for water leakage.
4. Does the tub have a hole in it? Is the tub rusted out?
5. Check the nut on the water inlet port. Is it tight?
6. Are the heater element nuts tight?
7. Check to be sure that all screws are tight. (motor and pump assembly, upper rack screws, etc.)

The dishwasher cycle will not advance (the lights will come on)

1. Is there voltage to the timer motor?
2. Are the cams in the timer rotating?
3. Is the dishwasher wired correctly?
4. Check to see if the timer is jammed.
5. Check to see if the selector switch is defective.

The dishwasher cycle will not advance other than at start

1. Is the thermostat functioning properly? This thermostat must be flush with the underside of the tub.
2. Check for correct timer settings.

The detergent cup won't open

1. Is there voltage to the solenoid?
2. Is the detergent mechanism adjusted properly?
3. Check to see if there is any binding at any point.
4. Check to see if any dishes are blocking the detergent cup door from opening.

The detergent cup won't close

1. Check the detergent cup actuator and cams.
2. Check to see if there is any binding at any point.
3. Is the detergent mechanism adjusted properly?

The main motor won't operate

1. Check for continuity at the start and run windings.
2. Is there voltage to the motor?
3. Is the motor jammed? Check for foreign debris in the pump assembly.
4. Check to see if there are any loose wiring terminals, or burned wires.
5. Check to see if the motor assembly is wired correctly.

The motor runs but goes into overload

1. Check the relay.
2. Is there any binding? Check the pump assembly for broken pieces of glass.
3. Be sure you have the correct voltage and the correct polarity.
4. Check to see if the motor windings are shorted.

Door liner hits side of tub in undercounter models

1. Check to see if the tub is square.
2. Check the installation, and correct as needed.

Dishwasher won't start

1. Is there voltage to the dishwasher? Check the plug, circuit breaker, or fuse box. Also, check the wires in the junction box, located behind the lower front panel.
2. Check the door switch for continuity of the switch contacts.
3. Check the selector switch for continuity of the switch contacts.
4. Check to see if the timer and selector switch has been wired correctly.
5. Check for any wires that might have come off the timer or switches.

The dishwasher repeats the cycle

1. Check the timer contacts; they should be open in the off position.
2. Check to see if the timer motor wires are shorted.

Wetting agent assembly leaks

1. Open the door assembly, and check for leaks in the holding tank.
2. Check the wetting agent assembly for proper operation.

Water siphons out through the drain pipe while the dishwasher is trying to fill

1. Is the drain line properly installed?
2. Check the installation instructions for proper installation.

COMMON WASHABILITY PROBLEMS

If there are no mechanical problems with the dishwasher's operation, and the complaints are that the dishwasher will not clean the dishes properly, or that the glassware is cloudy, etc., the next step will be to look at the best possible cause for the problem that the customer is experiencing with the dishwasher. Then, proceed to correct the problem. If necessary, instruct the user how to get better results from their dishwasher.

Poorly cleaned dishes

On occasion, there might be some food particles left on the dishes at the end of the cycle.

Possible cause: Water temperature might be too low. Remember, the water temperature should be 140–150 degrees F as it enters the dishwasher tub.
Solution: Check the water temperature at the closest faucet. Let the water from the hot water tap run before starting the dishwasher, in order to clear the water line of any cold water.

Possible cause: Not enough detergent for the degree of water hardness, or for the amount of dirty dishes to be cleaned.
Solution: Use one teaspoon of detergent for each grain of hardness, with three teaspoons of detergent as a minimum in soft water. The dishwasher will require extra detergent for greasy pans.

Possible cause: The detergent was placed in the wrong side of the dispenser cup.
Solution: Instruct the user how to fill the dispenser, and have the user reread the use and care manual so that the detergent is placed in the correct dispenser for the cycle that is selected.
Possible causes: Improper loading of the dishes into the dishwasher.
1. Blocking the spray nozzle. If a large bowl or pot is placed over the center of the lower rack, blocking the spray nozzle washing action when the lower rack is pushed in.
2. Larger items that shield smaller items from the washing action.
3. Observe to see if there is a nesting of the bowls, or silverware, so that the water cannot reach all surfaces.
4. The spray arms are blocked from turning. For example, tall items or an item that fell through the racks.

5. If an item blocks the detergent dispenser from opening, not allowing the detergent to mix with the water.

Solution: Instruct the user to reread the use and care manual for the proper instructions on how to load the dishes for proper cleaning.

Possible cause: Improper filling of water in the tub. Water pressure must be between 15 and 120 pounds per square inch. After the fill has stopped, check the water level in the tub. On most models, it should be even with the heating element in the bottom of the dishwasher.

Solution: If the water pressure is low, be sure that no other faucets are in use while the dishwasher is operating.

Possible cause: Not enough hot water.

Solution: Instruct the user to use the dishwasher when the hot water is not being used for laundry, baths, or showers.

Possible cause: If the dishwasher detergent is old and caked, it will not dissolve completely.

Solution: Instruct the customer to always purchase fresh detergent, and store the detergent in a dry place.

Etching

Etching occurs when the glass is pitted or eroded. It appears as a permanent film on the glass. The beginning stages of etching can be identified by an iridescent look, shades of blue, purple, brown, or pink when the glass if held at an angle to the light. In the advanced stages of etching, the glass surface appears frosted or cloudy.

Possible cause: Certain types of glass will etch, in any dishwasher, with the combination of soft water, the alkalinity of dishwasher detergents, and heat.

Solutions: There is no way to remove the filmy appearance caused by etching; the damage is permanent. There is no way to predict what glassware might be affected; it is not related to the cost or quality of the glass. To prevent etching from recurring:
1. Adjust the amount of detergent according to the water hardness.
2. Adjust the water temperature so that it enters the dishwasher at approximately 140 degrees F.
3. Recommend to the customer that the Energy Saver dry cycle be used.
4. Instruct the customer not to manually pre-rinse the dishes before loading them into the dishwasher.

Discoloration

Discoloration (red, black, or brown) of dishwasher interior or dishes

Possible cause: If iron or manganese is in the water, the dishes and/or the interior of the dishwasher might turn a red, black, or brown color.

Solution: A rust remover can be used to remove the discoloration from the dishwasher interior. With the dishwasher empty, turn to the rinse and hold cycle, and start the dishwasher. During the fill, open the door and add ½ cup of rust remover to the water. Allow the dishwasher to complete the cycle. Then, start the dishwasher on the normal wash cycle, with detergent, but without the dishes. To eliminate this condition from returning, the customer might have to install special filters to filter out the iron and manganese. Use a rust remover, as mentioned above, according to the manufacturer's recommendations, to remove the discoloration from the dishes and the glassware.

Lime deposits on the dishwasher interior

Possible cause: If there is a lot of calcium in the water, a lime film or deposit might eventually build up on the interior surfaces of the dishwasher.
Solution: You can try one of the following methods:
1. Use a mild scouring powder and a damp cloth to clean away the lime deposit.
2. With the dishwasher empty, turn the timer to the rinse and hold cycle. During the fill portion, open the door and add ½ cup of white vinegar to the water. Let the dishwasher complete the remaining cycle. Do not use detergent. After the cycle is completed, run the dishwasher with a regular load.
3. Use a product that removes lime deposits. Follow the manufacturer's directions.

Suds or foam in dishwasher

Possible causes:
1. Sudsing in the dishwasher is caused by protein foods (milk, eggs, etc.), and an insufficient amount of detergent.
2. The water in the dishwasher is not hot enough to activate the defoaming agents in the dishwasher detergent.
3. The user has used a non-automatic dishwasher detergent.
Solution: Increase the amount of detergent to reduce the sudsing. Dishwasher detergents contain defoaming agents to break down the suds in the dishwasher water. Check to be sure that the water temperature is between 140 and 150 degrees F. Only use detergents that are for automatic dishwashers.

Darkened aluminum

Possible cause: A combination of water, heat, and alkaline foods will darken or stain aluminum products.
Solution: To remove this discoloration, instruct the customer to use an aluminum cleaner, and to clean the item by hand. Never allow undissolved dishwasher detergent to come in direct contact with the metal. Avoid placing aluminum items in the lower rack, right in front of the detergent dispenser.

Discoloration of copper

Possible cause: Some copper items will discolor when washed by hand, as well as when washed in the dishwasher, because of the heat and detergent alkalinity.
Solution: Instruct the customer to use a copper cleaner to restore the copper color.

Cracking (crazing) of china

Possible cause: Crazing is the appearance of tiny cracks that appear over the entire surface of the china. It can occur when porous earthenware, good china that is very old, or lower quality china, is exposed to heat and moisture.

Solution: Once the glaze is cracked or crazed, the damage is permanent. This characteristic is inherent in some clayware; and this type of damage can occur during use, handwashing, or automatic dishwashing.

Chipping of china and crystal

Possible cause: Chipping usually occurs during normal use and handling, and simply might not be noticed until the dishes are removed from the dishwasher. When the dishes are loaded into the dishwasher according to the manufacturer's instructions, there is nothing in the dishwasher that can chip the dishes. The dishes should only come in contact with the cushioned vinyl coated racks.

Solution: Instruct the user to follow the manufacturer's instructions, in the use and care manual, for loading the dishwasher properly.

Metal marks on dishes and glassware

Possible cause: If a metal item, especially aluminum, touches a dish in the dishwasher, a metal mark might result. This symptom appears as small black or gray marks, or streaks, on dishes or glasses. However, most metal marks occur during normal use, when the dishes come in contact with the flatware.

Solution: Instruct the user to load the dishwasher carefully, in order to prevent metal items from touching other dishes. There are products on the market that will remove these marks. Have the user read over the use and care manual, or check with the manufacturer for this information.

Staining of melamine dinnerware

Possible cause: Melamine dinnerware stains from contact with coffee, tea, and some fruit juices. If the surface is worn, it will stain more readily.

Solution: Some specialty products on the market are recommended for removing these stains. Instruct the customer to read over the use and care manual, or check with the manufacturer for this information.

Melting or warping of plastic items

Possible cause: Some plastic items cannot be exposed to the temperatures usually found in dishwashers without changing shape.

Solution: Once the plastic item has distorted, it cannot be returned to its original shape. In order to minimize, or prevent, plastic items from warping or melting, instruct the user to do one of the following:

1. Choose the air dry cycle to dry the dishes.
2. Place the plastic items on the top rack.
3. Purchase and use plastic items that are labeled "dishwasher-proof".

Discoloration of silverplate

Possible cause: When silverplate takes on a copper or bronze colored appearance, the silverplate has worn thin and the base metal is showing through. The combination of dishwasher detergent and the lack of hand toweling might result in discoloration of this base metal.

Solution: This discoloration can usually be removed by polishing the item with a silver polish, or by soaking the item in vinegar for about 10 minutes. This is only a temporary solution, however. Only a replating with silver, by a jeweler, will correct the problem.

Tarnishing of silverware (sterling or silverplate)

Possible cause: Sulfur in the water supply might be the cause. This effect might be accelerated by the automatic dishwasher because of the higher water temperature, and because the usual hand drying with a towel has been eliminated.

Solution: Because sulfur cannot be readily removed from the water supply, frequent polishing is the only answer.

Possible cause: Silver will tarnish easily if it is left in contact with foods such as mayonnaise and eggs.

Solution: If silverware has been in contact with such foods, instruct the user to rinse the item thoroughly as soon as possible.

Bluish discoloration of stainless steel

Possible cause: A bluish discoloration of some types of stainless steel is caused by heat and the alkalinity of the automatic dishwasher detergent.

Solution: This discoloration can be removed by using a paste of baking soda and water, or a stainless steel cleaner.

Corrosion or rusting of stainless steel

Possible cause: When the protective oxide film on the surface of the steel is removed, corrosion will take place as with ordinary steels. Certain foods will remove the oxide film. They are: table salt, vinegar, salad dressings, milk and milk products, fruits and juices, tomatoes and tomato products, and butter. However, if the stainless is washed, rinsed, and dried thoroughly, the oxygen of the air will heal the breaks in the oxide film and return the stainless property to the steel. But, if food is not washed off promptly, the air cannot heal the break, and corrosion will occur.

Solution: There is no permanent solution. To minimize rusting, instruct the user to rinse or wash the flatware as soon as possible after use.

DISHWASHER MAINTENANCE

The dishwasher's interior is normally self-cleaning. However, there are times when the customer will have to remove food particles or broken glass in the bottom of the tub. Inform the customer to clean the bottom edge of the dishwasher tub, which is sealed off by the gasket when the door is closed. Food and liquids drip onto this area when the dishwasher is loaded. The control panel should be cleaned with a soft damp cloth. Tell the customer not to use any abrasive powders or cleaning pads. Also, advise the customer to read the use and care manual for proper maintenance procedures on their brand of dishwasher. Dishwashers are designed to flush away all normal food soils that have been removed from the dishes. However, on occasion, certain foreign objects, such as fruit pits, bottle caps, etc., might collect in the openings of the pump, and these items should be removed periodically to avoid clogging the drain system. Also, on occasion, some of these foreign objects can get caught in the spray arm openings, and will have to be cleaned out. Check the racks carefully to see if there are any nicks or cuts in the vinyl. These nicks and cuts can be repaired. A liquid vinyl repair material is available through the manufacturer, or any appliance supply store.

REPAIR PROCEDURE

Each of the following repair procedures is a complete inspection and repair process for a single dishwasher component; containing the information you need to test a component that might be faulty, and to replace it, if necessary.

Water valve

The typical complaints associated with water valve (Fig. 7-5) failure are: 1. The dishwasher will run, but no water will enter dishwasher. 2. The dishwasher will overfill, and leak onto the floor. 3. When the dishwasher is off, water still enters the tub. 4. The dishes are not clean, not enough water enters the tub.

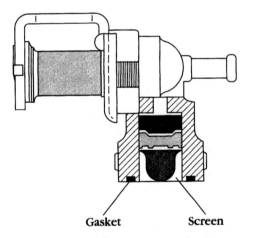
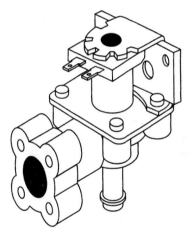

Gasket   Screen

7-5 A typical dishwasher water valve.

1. *Verify the complaint* Verify the complaint by operating the dishwasher through its cycles. Listen carefully, and you will hear if the water is entering the dishwasher.
2. *Check for external factors* You must check for external factors not associated with the appliance. Is the appliance installed properly? Is the voltage supply correct for the dishwasher?
3. *Disconnect the electricity* Before working on the dishwasher, disconnect the electricity. This can be done by pulling the plug out of the wall receptacle.

Be sure that you only remove the dishwasher plug. Double check to ensure that the electrical supply has been disconnected before removing any service panels. Another way to disconnect the electricity is at the fuse panel, or the circuit breaker panel. Turn off the electricity.

4. *Remove the bottom panel* In order to gain access to the water valve, the bottom panel must be removed (Fig. 7-6). The bottom panel is held on with 2 or 4 screws, depending on the model. Remove the screws, and remove the panel.

Bottom screws

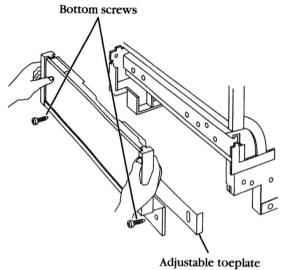

7-6 Removing the bottom panel.

Adjustable toeplate

5. *Remove the wire leads* In order to check the solenoid coil on the water valve, remove the wire leads that connect to the solenoid coil from the wire harness (Fig. 7-7). These are slide-on terminal connectors attached to the ends of the wire. Just pull them off.

6. *Test the water valve* Using the ohmmeter, set the range on R × 1000, and place the probes on the solenoid coil terminals (Fig. 7-8). The meter should read between 700- and 900-ohms resistance. If not, replace the water valve. If you determine that the water valve is good, but there is little water flow through the valve, inspect the inlet screen. If this screen is filled with debris, it must be cleaned out. To accomplish this, use a small flat blade screwdriver and pry out the screen (Fig. 7-9). Then, wash out the screen, being sure that all of the debris is removed. Reinstall the screen. Turn on the water supply. Plug in the dishwasher. Allow the water to enter the tub to check the flow rate of the water valve. The tub must be empty because this will allow you to check the flow rate properly. On a normal fill, the water line should be over the heating element within two minutes (Fig. 7-10). When you turn the dishwasher on and energize the water valve, but no water enters the dishwasher tub, then replace the water valve. If the water valve checks okay, then check the timer and the wiring harness.

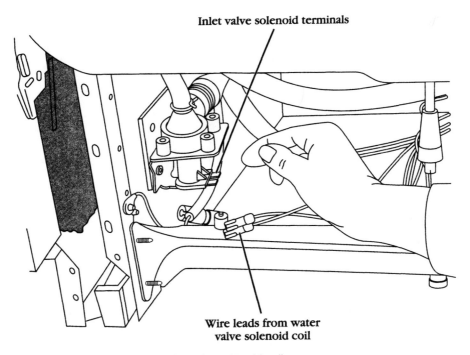

Inlet valve solenoid terminals

Wire leads from water
valve solenoid coil

7-7 Removing the wire leads from the solenoid coil.

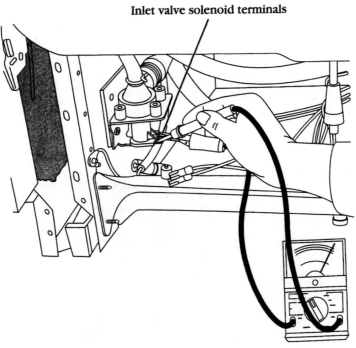

Inlet valve solenoid terminals

7-8 Connect meter probes to the water valve solenoid coil.

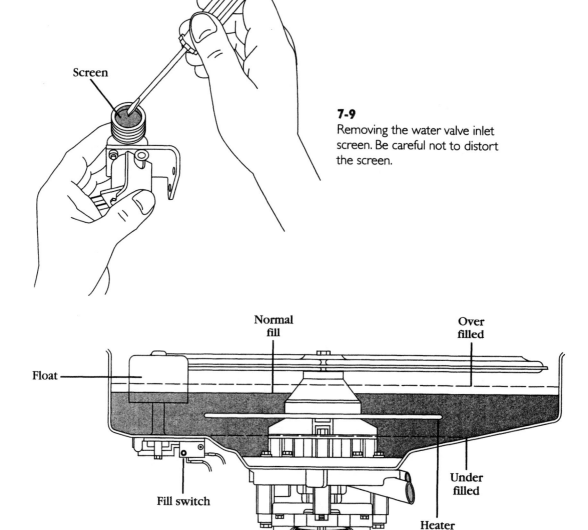

7-9
Removing the water valve inlet screen. Be careful not to distort the screen.

Screen

Normal fill

Over filled

Float

Under filled

Fill switch

Heater

7-10 Diagram indicates water level in dishwasher tub.

7. *Remove the water valve* Before removing the water valve, turn off the water supply to the dishwasher. Then, disconnect the water supply line from the inlet end of the water valve, and remove the fill hose from the outlet side (Fig. 7-11). Next, remove the screws that hold the water valve to the chassis

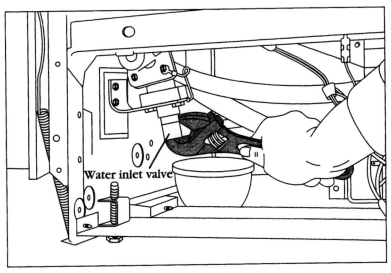

7-11 Disconnect water supply line.

of the dishwasher. To install the new water valve just reverse the disassembly procedure, and reassemble. Check for water leaks. If none are found, reinstall the bottom panel, and restore the electricity to the dishwasher.

8. *Operate the dishwasher* Set the timer and control settings to operate the dishwasher through its cycle.

Motor and pump assembly

The typical complaints associated with pump and motor failure are: 1. The water will not drain out of the dishwasher. 2. Poor washability of the dishes. 3. When the motor runs, there are loud noises. 4. The fuses or circuit breaker will trip when you start the dishwasher. 5. Water leaking from bottom of the dishwasher, or water leaking on the motor/pump assembly.

1. *Verify the complaint* Verify the complaint by operating the dishwasher through its cycles. Listen carefully, and you will hear if there are any unusual noises, or if the circuit breaker trips.

2. *Check for external factors* You must check for external factors not associated with the appliance. Is the appliance installed properly? Does the appliance have the correct voltage?

3. *Disconnect the electricity* Before working on the dishwasher, disconnect the electricity. This can be done by pulling the plug out of the wall receptacle. Be sure that you only remove the dishwasher plug. Or disconnect the electricity at the fuse panel or at the circuit breaker panel. Turn off the electricity.

4. *Remove the bottom panel* In order to gain access to the pump and motor assembly, the bottom panel must be removed (Fig. 7-6). The bottom panel is held on with 2 or 4 screws, depending upon the model. Remove the screws and remove the panel.

5. *Disconnect the motor wire leads* Disconnect the motor wire leads from wiring harness. Check the motor windings for continuity (Fig. 7-12). Check for resistance from the common wire lead to the run winding (Fig. 7-13). Then, check the resistance from the common wire lead to the drain winding, and the common wire lead to the wash winding (Fig. 7-14). To check for a grounded winding in the motor, take the ohmmeter probes and check from each motor wire lead terminal to the motor housing (Fig. 7-15). The ohmmeter will indicate continuity if the windings are grounded (Fig. 7-16). If the motor/pump assembly showed signs of water leaking, then, replace the motor/pump assembly as a complete unit. Most part manufacturers give a one year warranty on the motor/pump assembly. There is no advantage in tearing down the motor/pump assembly; to only replace 1 or 2 parts; and receiving only a partial warranty. Replacing the motor/pump assembly as a complete assembly will save you time and money in the long run. If the motor/pump assembly checks okay, then check the timer and the motor relay (if the model you are repairing has one); and check for a kinked or plugged drain line.

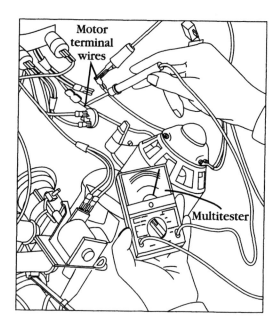

7-12
Check the motor windings for continuity.

6. *Remove motor and pump assembly* Remove the lower dishwasher rack from the dishwasher. As shown in Fig. 7-17, remove the wash tower and spray arm assembly. If there is a filter, remove it also. Remove the motor wiring leads from the wiring harness, then, remove the drain line from the pump assembly. Reach underneath the tub and rotate the four pump hold downs 90 degrees inward (Fig. 7-18). You are now ready to remove the motor/pump assembly. Lift the motor/pump assembly out from the inside of the tub (Fig. 7-19). Keep the work area dry to help prevent electrical shocks.

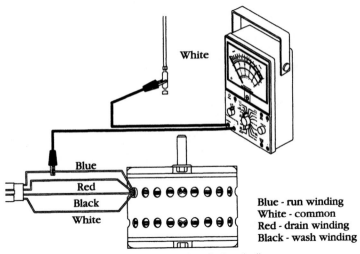

Blue - run winding
White - common
Red - drain winding
Black - wash winding

7-13 Check from the common to the drain winding.

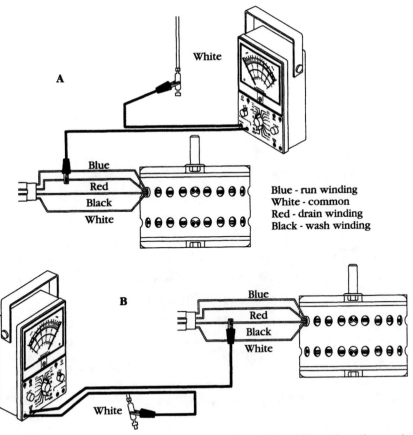

Blue - run winding
White - common
Red - drain winding
Black - wash winding

7-14 Check from the common to the drain winding (A), and to the wash winding (B).

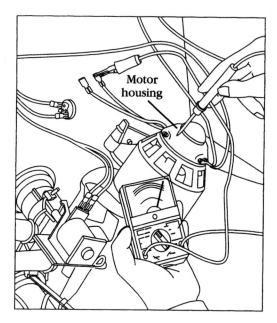

7-15
Check for the grounded motor.

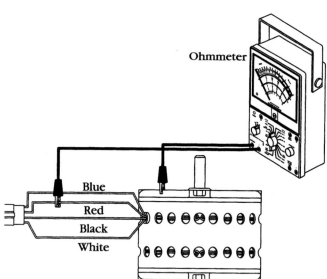

Ohmmeter

Blue
Red
Black
White

7-16 The grounded motor.

7. *Reinstall motor and pump assembly* To reinstall the motor/pump assembly, just reverse the order of step 6. Before restoring the electricity to the dishwasher, pour a gallon of water into the tub and check for leaks underneath the tub. Restore the electricity to the dishwasher, and run the dishwasher through a cycle. Check for leaks again. If no leaks are found, reinstall the bottom panel.

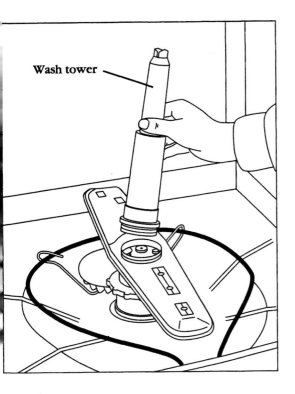

Wash tower

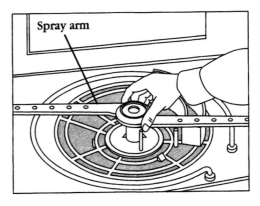

Spray arm

7-17
Remove the wash tower and
the spray arm assembly.

Dishwasher timer

The typical complaints associated with dishwasher timer failure are: 1. The cycle will
not advance. 2. The dishwasher won't run at all. 3. The dishwasher will not fill. 4.
The dishwasher will not pump the water out.

1. *Verify the complaint* Verify the complaint by operating the dishwasher
   through its cycles. Before you change the timer, check the other
   components controlled by the timer.
2. *Check for external factors* You must check for external factors not
   associated with the appliance. Is the appliance installed properly? Does the
   appliance have the correct voltage?
3. *Disconnect the electricity* Before working on the dishwasher, disconnect the
   electricity. This can be done by pulling the plug out of the wall receptacle.
   Be sure that you only remove the dishwasher plug. Or disconnect the
   electricity at the fuse panel or circuit breaker panel. Turn off the electricity.
4. *Remove the console to gain access* Begin with removing the four screws
   from the console to access the timer (Fig. 7-20). Turn the timer knob
   counterclockwise to remove it from the timer shaft and slide the indicator
   dial off the shaft. Remove the console panel from the dishwasher. On some
   models, the latch handle knob will also have to be removed.
5. *Test the timer* Disconnect the timer motor wire leads from the timer
   assembly. Using the ohmmeter, set the range on R × 1000, and place the
   probes on the timer motor terminals (Fig. 7-21). The meter should indicate

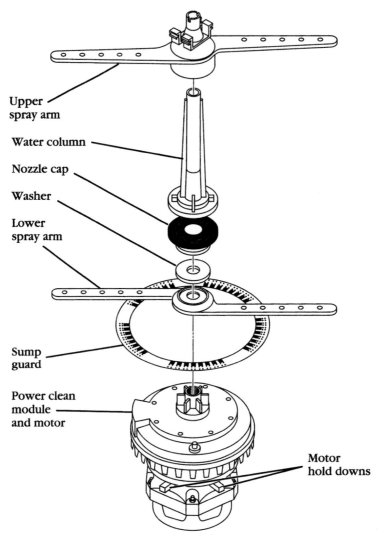

Upper spray arm

Water column

Nozzle cap

Washer

Lower spray arm

Sump guard

Power clean module and motor

Motor hold downs

7-18 An exploded view of a dishwasher motor/pump assembly.

some resistance. If not, replace the timer. If the motor checks out, then check the door latch switch, cycle extender relay, and the float switch assembly.

6. *Remove the timer* To remove the timer, remove the timer mounting screws (Fig. 7-22). Remove the wire lead terminals from the timer. Mark the wires as to their location on the timer. Some timers have a disconnect terminal block instead of individual wires, which makes it easier to remove the timer wires.

7. *Install the new timer* To install a new timer, just reverse the disassembly procedure, and reassemble. Reinstall the console panel, and restore the electricity to the dishwasher. Test the dishwasher operation.

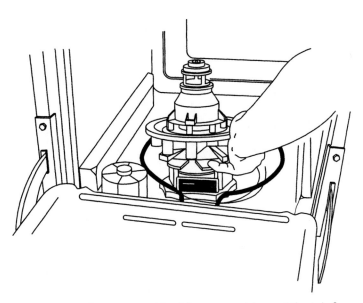

7-19 Remove the motor/pump assembly. After removal, inspect the tub for rust.

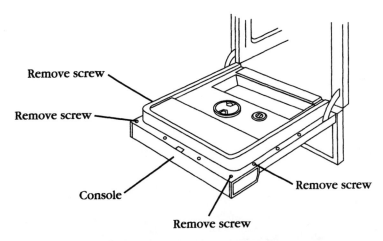

Remove screw

Remove screw

Console

Remove screw

Remove screw

7-20 Remove the screws to access the control panel.

The dishwasher door is hard to close

The typical complaints associated with dishwasher door being hard to close are: 1. The door is hard to close. 2. The door won't latch. 3. The dishwasher won't run.

1. *Verify the complaint* Verify the complaint by trying to close and latch the door.
2. *Check for external factors* You must check for external factors not associated with the appliance. Is the appliance installed properly? Be sure that the dishwasher door is not binding against the side cabinet.

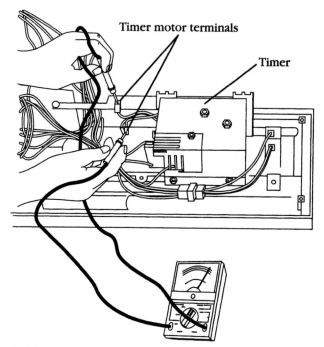

7-21 Attach the meter probes and test for continuity.

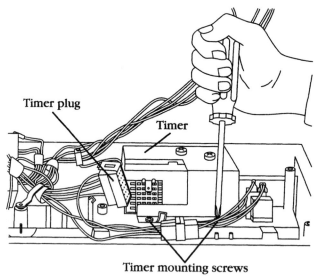

7-22 Removing the timer.

3. *Run the dishwasher* By pushing on the door, you finally latch the door closed. Next, run the hot water closest to the dishwasher to flush out the supply line. Then, turn the timer to the normal wash cycle with heated dry, and let it run through the entire cycle. By doing this procedure, the gasket will soften enough to form itself to the tub, while it is compressed. This should make it easier to latch the door. If not, adjust the latch assembly. If it is still too hard to close, then replace the door gasket.

4. *Replace the door gasket* To remove the old gasket, remove the screws or clips that hold the old gasket in place (Fig. 7-23). On some models, the gasket is pressed into the inner door assembly. Remove the inner panel from the dishwasher, and pull off the gasket from the panel (Fig. 7-24). Soak the new gasket in warm water to make it more flexible.

5. *Test a new gasket* After you have installed the new door gasket, close the dishwasher door, and test to see if the door latches without pushing hard against it.

6. *Check for water leaks* Run the dishwasher through another cycle to check for water leaking out the door. Most models have a baffle, or tub, gasket. If the dishwasher that you're working on has one, then check to make sure it is not defective. The tub gasket is located either on the tub, or behind the inner door panel.

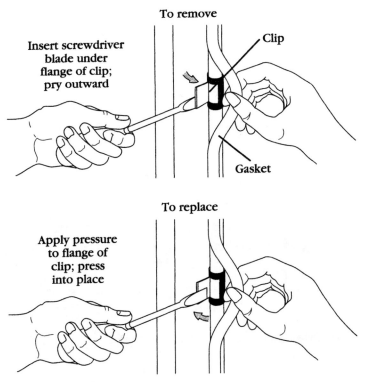

7-23 Removing dishwasher door gasket.

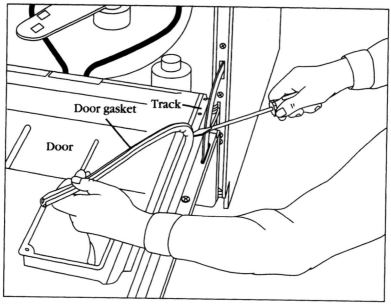

7-24 Pull off the gasket. Be careful not to damage the door liner.

Dishwasher heating element

The typical complaints associated with the heating element are: 1. Dishes are still wet at the end of the cycle. 2. The dishwasher stalls in the middle of the cycle.

1. *Verify the complaint* Verify the complaint by operating the dishwasher starting at the dry cycle.

2. *Check for external factors* You must check for external factors not associated with the appliance. For example, check the energy saver switch. Is it set for heat dry? Is the appliance installed properly? Does the appliance have the correct voltage?

3. *Disconnect the electricity* Before working on the dishwasher, disconnect the electricity. This can be done by pulling the plug out of the wall receptacle. Be sure that you only remove the dishwasher plug. Or disconnect the electricity at the fuse panel, or the circuit breaker panel. Turn off the electricity.

4. *Remove the bottom panel* In order to gain access to the heater wire terminals, the bottom panel must be removed (Fig. 7-6). The bottom panel is held on with 2 or 4 screws, depending on the model. Remove the screws and remove the panel.

5. *Test the heating element* To test the heating element (Fig. 7-25), remove the wires from the heating element terminals (Fig. 7-26). These are slide on terminal connectors attached to the ends of the wire. Use the ohmmeter to check for continuity between the two element terminals (Fig. 7-27). If the meter indicates no continuity between the terminal ends, replace the heater. To check for a shorted heating element, take one end of the ohmmeter probe and touch the element terminal; then, with the other probe, touch the

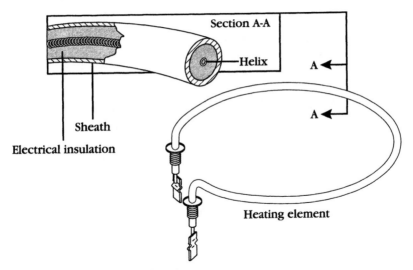

7-25 The dishwasher heating element.

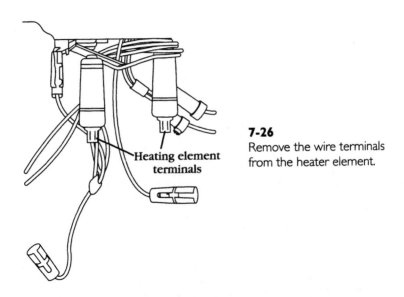

7-26
Remove the wire terminals
from the heater element.

sheath (outer cover of element) (Fig. 7-28). If the meter indicates continuity, the element is shorted, and should be replaced.

6. *Remove the heating element* To remove the heating element (with the wires already removed from the heater terminals), unscrew the locknuts that hold the element in place. From inside the tub, remove the heating element (Fig. 7-29).

7. *Install a new heating element* To install a new element, just reverse the disassembly procedure and reassemble. Then, test the new element by repeating step 1.

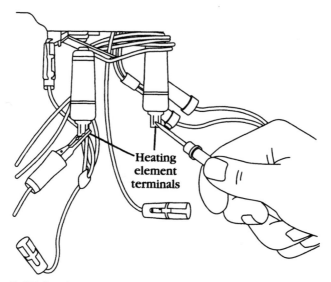

7-27 Set the meter on the ohm's scale. Connect the probes to the heater terminals.

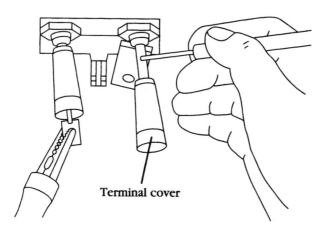

Terminal cover

7-28 To check for a shorted element, attach the meter probe to one terminal and attach the other probe to the sheath.

Cycle selector switch

The typical complaints associated with the cycle selector switch are: 1. Inability to select a different cycle. 2. The consumer inadvertently selected the wrong cycle.

1. *Verify the complaint* In Fig. 7-30, verify the complaint by trying to select different cycles.
2. *Check for external factors* You must check for external factors not associated with the appliance. For example, is there any physical damage to the component?

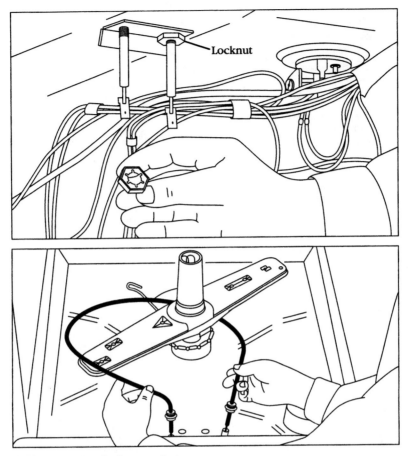

7-29 Removing the heater element.

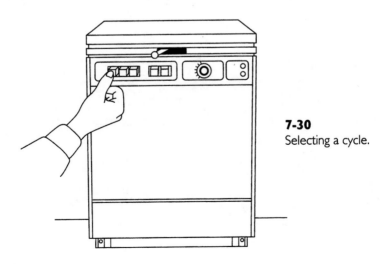

7-30
Selecting a cycle.

3. *Disconnect the electricity* Before working on the dishwasher, disconnect the electricity. This can be done by pulling the plug out of the wall receptacle. Be sure that you only remove the dishwasher plug. Or disconnect the electricity at the fuse panel or circuit breaker panel. Turn off the electricity.

4. *Remove the control panel* To remove the control panel, remove the screws along the top inside edge of the door (Fig. 7-20). On some models, you might have to remove the door latch knob, in order to completely remove the control panel.

5. *Test the cycle selector switch* To test the cycle selector switch, remove all wires from the switch. Just remember, you will have to identify the wires according to the wiring diagram, in order to reinstall them back onto the cycle selector switch properly. Take your ohmmeter, and check for continuity on the switch contacts; press that switch which coincides to the terminals that are being checked (Fig. 7-31). At this point, you have to use the wiring diagram to identify the switch contacts.

6. *Remove the cycle selector switch* To remove the cycle selector switch, remove the screws that hold the component to the control panel (Fig. 7-32).

7. *Reinstall the cycle selector switch* To reinstall the cycle selector switch, just reverse the disassembly procedure, and reassemble. Remember, you will have to identify the wires according to the wiring diagram in order to reinstall them back onto the cycle selector switch properly.

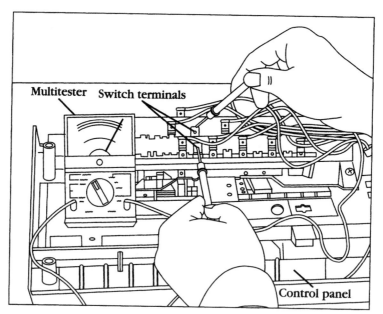

7-31 Testing the selector switch contacts.

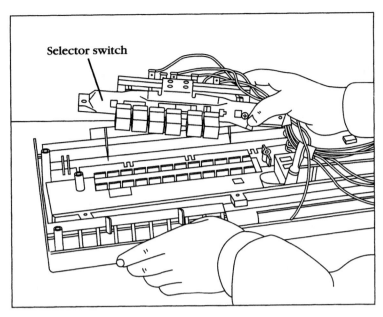

7-32 Removing the selector switch.

Dishwasher door switch

Sometimes the dishwasher door switch malfunctions and the door will not close properly.

1. *Verify the complaint* In Fig. 7-33, verify the complaint by closing the dishwasher door, and turning the timer dial to start the wash cycle.
2. *Check for external factors* You must check for external factors not associated with the appliance. Is the appliance installed properly? Does the appliance have the correct voltage? Is there any physical damage to the component?
3. *Disconnect the electricity* Before working on the dishwasher, disconnect the electricity. This can be done by pulling the plug out of the wall receptacle. Be sure that you only remove the dishwasher plug. Or disconnect the electricity at the fuse panel or circuit breaker panel. Turn off the electricity.
4. *Remove the control panel* To remove the control panel, remove the screws along the top inside edge of the door (Fig. 7-20). On some models, you might have to remove the door latch knob in order to completely remove the control panel.
5. *Test the door switch* To test the door switch, remove the two wires from the switch (Fig. 7-34). Close and latch the door. With your ohmmeter, check for continuity between the two terminals on the switch. Then, open the door latch and check for no continuity between the terminals. If the switch fails these tests, replace the switch.

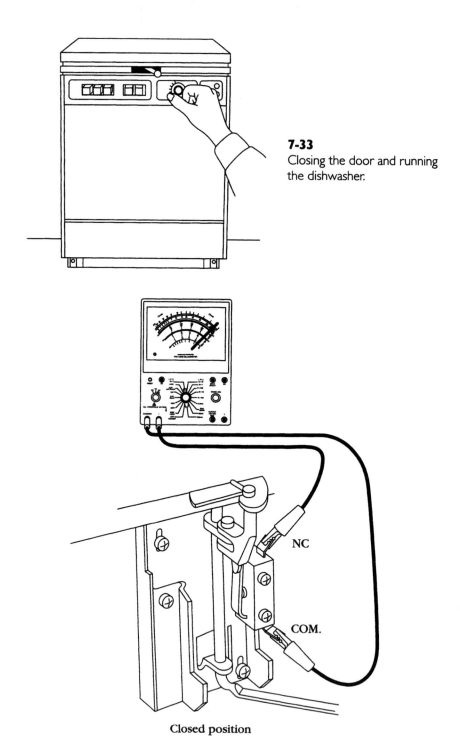

7-33
Closing the door and running the dishwasher.

NC

COM.

Closed position

7-34 The dishwasher latch assembly in closed position.

6. *Remove the door switch* Remove the screws that secure the switch in place. Remove the switch.

7. *Install new door switch* To install the new door switch, just reverse the disassembly procedure, and reassemble. Then, reconnect the electricity, and test the dishwasher.

Float and float switch

Sometimes the dishwasher float and the float switch malfunctions and water will not enter the tub, or water overfills and spills onto the floor.

1. *Verify the complaint* Verify the complaint by closing the dishwasher door, and turning the timer dial to start the wash cycle.

2. *Check for external factors* You must check for external factors not associated with the appliance, as was done in the previous sections.

3. *Disconnect the electricity* Before working on the dishwasher, disconnect the electricity. This can be done by pulling the plug out of the wall receptacle. Be sure that you only remove the dishwasher plug. Or disconnect the electricity at the fuse panel, or the circuit breaker panel. Turn off the electricity.

4. *Check the float* Check the float from the inside of the tub (Fig. 7-35). Be sure that the float moves freely up and down. If there is any soap build-up around the float, clean it off.

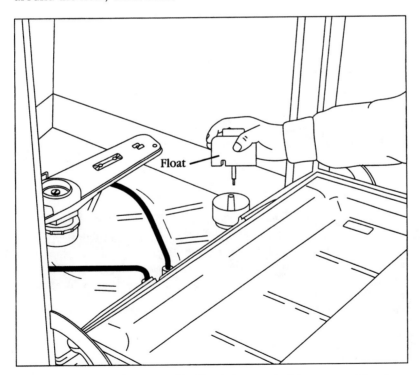

7-35 Inspecting the float assembly. Check for soap build-up around the stem and inside of the float.

5. *Remove the bottom panel* In order to gain access to the float switch terminals, the bottom panel must be removed (Fig. 7-6). The bottom panel is held on with 2 or 4 screws, depending on the model. Remove the screws, and remove the panel.

6. *Test the float switch* The float switch is located under the float, and underneath the tub (Fig. 7-36). Remove the wires from the terminals, and test for continuity. Lift the float, and there should be no continuity. Let the float rest, and you should have continuity. If the test fails, replace the float switch.

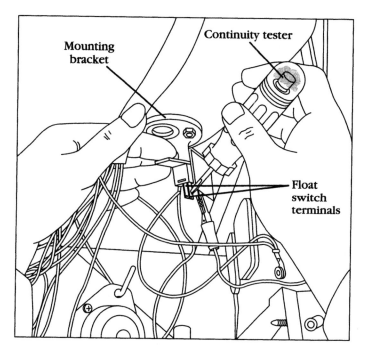

7-36 Testing the float switch contacts.

7. *Remove the float switch* To remove the float switch, you will have to remove the screws that hold the switch in place (Fig. 7-37).

8. *Install a new float switch* To install the new float switch, just reverse the disassembly procedure, and reassemble. Then, reconnect the electricity, and test a dishwasher cycle that fills the dishwasher.

DIAGNOSTIC CHARTS

The following diagnostic charts will help you to pinpoint the likely causes of the various dishwasher problems (Figs. 7-38 through 7-44). Schematics (Figs.7-45 through 7-47) start on page 148.

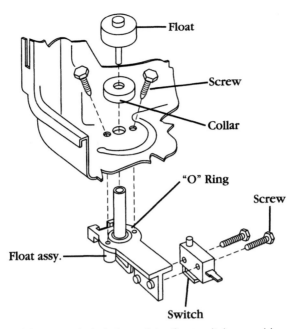

7-37 An exploded view of the float switch assembly.

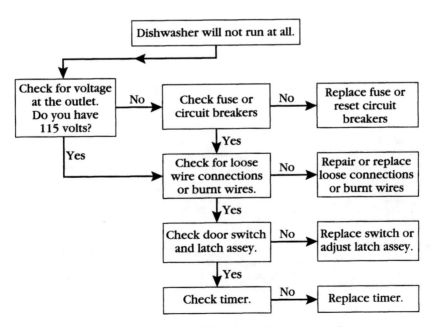

7-38 Dishwasher diagnostic flow chart: dishwasher will not run at all.

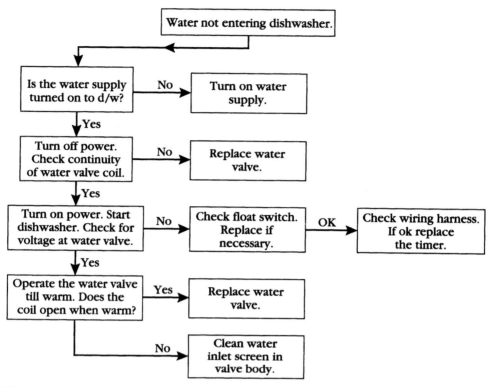

7-39 Dishwasher diagnostic flow chart: water not entering the dishwasher.

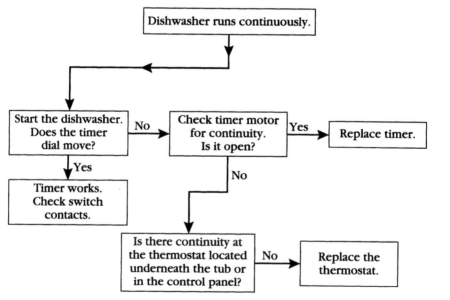

7-40 Dishwasher diagnostic flow chart: dishwasher runs continuously.

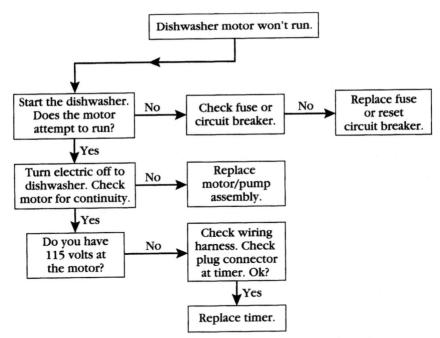

7-41 Dishwasher diagnostic flow chart: dishwasher motor won't run.

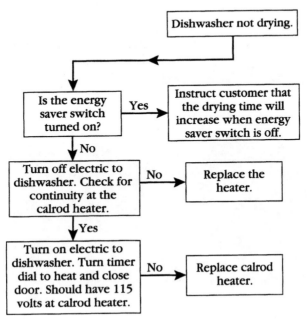

7-42 Dishwasher diagnostic flow chart: dishwasher won't dry.

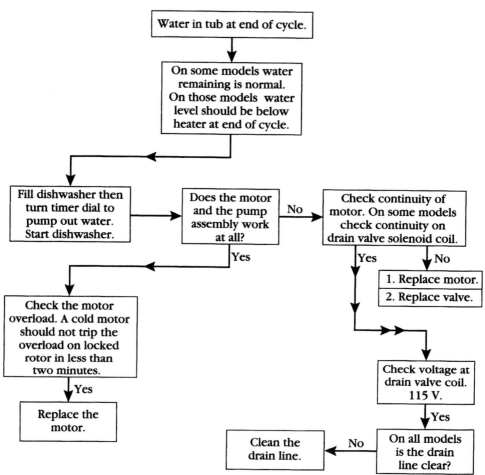

7-43 Dishwasher diagnostic flow chart: water in the tub at the end of cycle.

Water valve noise chart

Complaint				
Noise	Description	When noise occurs	Which water valve parts	External factors
Whistle	Makes a "hissing" to a shrill whistle.	While the water valve is open.	Outlet assembly and hose to fill funnel, flow-washer.	* Close or open the supply water valve to change the pressure band. * Water heater temperature too high. Set the water heater thermostat to 140 degrees Fahrenheit.
Flutter	Low frequency "rumble."	When the water valve is about to close.	Flow-washer.	
Chatter	Low frequency pulsing.	When the water valve closes.	Spring, diaphragm.	* Long runs of unsupported pipes. You must secure the pipes to stop any vibrations. * Water supply line to the water valve is undersized. Must have 3/8" I.P.S. or 1/4" I.D. * Water heater temperature is too high. Set the water heater thermostat at 150 degrees Fahrenheit maximum.
Water hammer	The water valve will make a single "thud."	At closing.	Armature spring, diaphragm.	

7-44 Water valve noise chart.

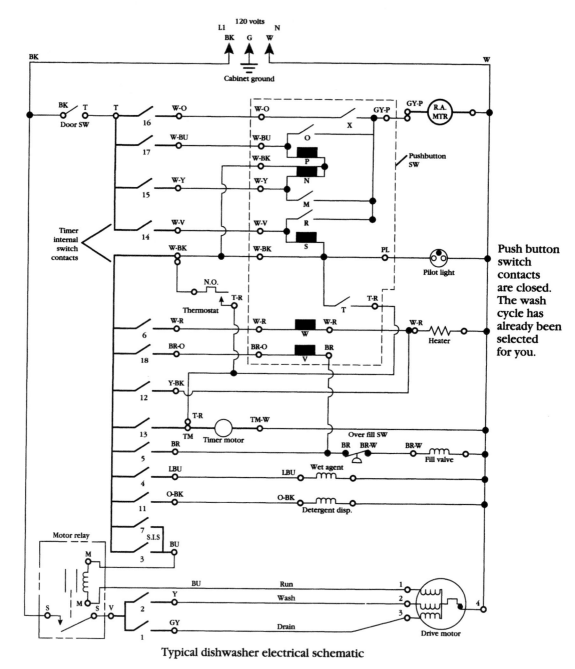

Typical dishwasher electrical schematic

7-45 The typical dishwasher electrical schematic. The wash cycle has already been selected. The pushbutton switch contacts are closed.

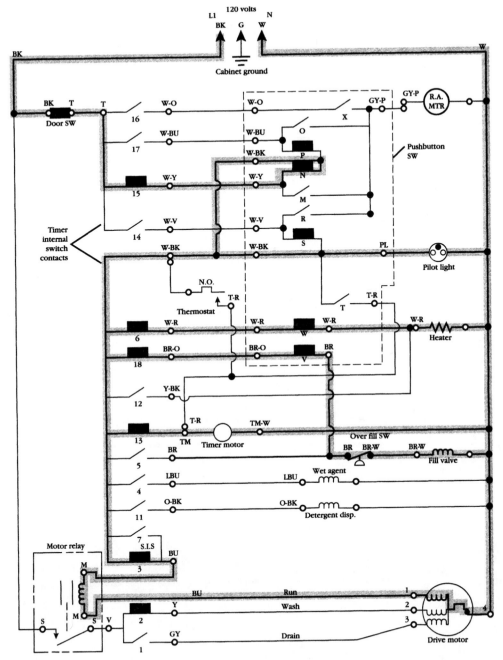

7-46 This illustration shows the active circuits, while the dishwasher tub is filling up with water.

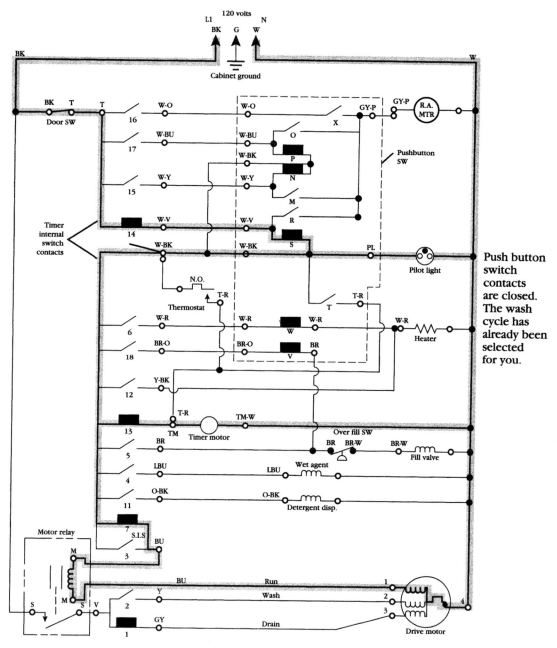

7-47 This illustration shows the active circuits, while the dishwasher is draining the water out of the tub.

8
Garbage disposers

The garbage disposer provides a convenient and sanitary way to dispose of food waste (Fig. 8-1). No sorting or separating of the food waste is necessary. The garbage disposer is designed to handle all types of food waste. If, however, the garbage disposer drains into a septic tank, inorganic wastes, such as egg shells, lobster, crab, and shrimp shells should be kept at a minimum. Items such as tin cans, glass, china, bottle caps, metal, etc., should never be placed in the garbage disposer because they might damage the appliance and plug the drain.

TYPES

There are two types of operating methods available in garbage disposers.

Continuous feed disposers

A continuous feed disposer requires an on-off electrical switch that is remote from the disposer. Before food is placed into the hopper, the user must turn on the cold water faucet, and turn on the electrical switch. Then, the food waste is placed into the hopper. The food waste will be ground up, and expelled into the drain.

Batch feed disposers

In a batch feed disposer, the on-off operation of the disposer is controlled by the stopper. The on-off switch is built into the hopper. The stopper is the component that completes the circuit that allows the disposer motor to run. The hopper will hold the food waste until the user is ready to dispose of the waste. The food waste will be ground up, and expelled into the drain.

PRINCIPLES OF OPERATION

When the food is placed into the hopper, and the disposer is running, the centrifugal force throws the food waste outward against the cutting edges of the shredder ring (Fig. 8-2). The pivoting impeller arms, attached to the flywheel, push the food waste around and into the teeth of the shredder ring (Fig. 8-3). As the food waste is pushed against, and cut by, the shredder ring, water running into the hopper flushes the ground food waste between the flywheel and the shredder ring; then washing into the drain housing assembly, where it is expelled into the drain.

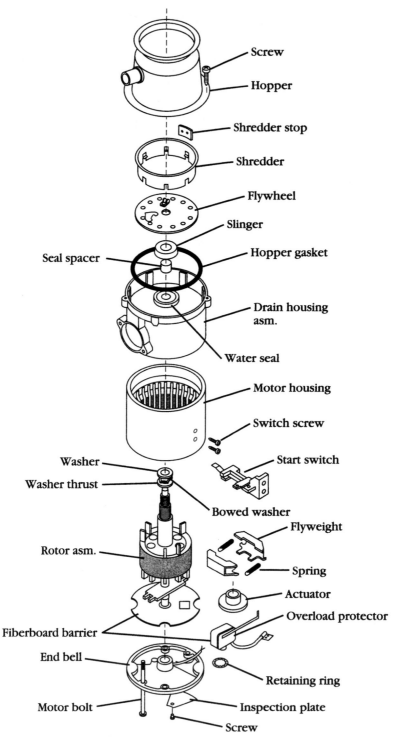

Screw

Hopper

Shredder stop

Shredder

Flywheel

Slinger

Seal spacer

Hopper gasket

Drain housing
asm.

Water seal

Motor housing

Switch screw

Washer

Start switch

Washer thrust

Bowed washer

Flyweight

Rotor asm.

Spring

Actuator

Overload protector

Fiberboard barrier

End bell

Retaining ring

Motor bolt

Inspection plate

Screw

8-1 Exploded view of a garbage disposer.

8-2
Food waste is forced outward against shredder ring.

8-3
Impeller directing food waste into teeth of the shredder ring.

SAFETY FIRST

Any person who cannot use basic tools, or follow written instructions, should *not* attempt to install, maintain, or repair any garbage disposers. Any improper installation, preventive maintenance, or repairs could create a risk of personal injury or property damage.

If you do not fully understand the installation, preventive maintenance, or repair procedures in this chapter, or if you doubt your ability to complete the task on your garbage disposer, then please call your service manager.

These precautions should also be followed:

1. Do not put your fingers or hands into the garbage disposer.
2. When removing foreign objects, always use a long-handled pair of tongs or pliers.

3. To reduce the risk of flying debris that can be expelled by the disposer, always be sure that the splash guard is properly installed.
4. When replacing the garbage disposer, always be sure that the disposer is properly grounded.
5. Before attempting to free a jam in the garbage disposer, always disconnect the electric supply to the disposer.

GARBAGE DISPOSERS IN GENERAL

Much of the troubleshooting information in this chapter covers the various types of garbage disposers in general, rather than specific models, in order to present a broad overview of service techniques. The pictures and illustrations that are used in this chapter are for demonstration purposes; for clarifying the description of how to service these units; and in no way reflects on a particular brand's reliability.

To free jams from foreign objects

When foreign material falls into the disposer, and the motor jams, it will trip the motor overload protector; otherwise known as the reset button (Fig. 8-1). When this happens, follow these steps to free the jam:
1. Turn off the electricity to the disposer, and shut off the water.
2. Insert the service wrench into the center hole at the bottom of the disposer (Fig. 8-4), and work the wrench back and forth until the motor assembly is turning freely.
3. To free a tight jam, insert a prying tool alongside the grinding protrusion near the outside edge of the flywheel. Be sure that you place the prying tool on the proper side of the protrusion so that when pressure is applied, the flywheel will move (Fig. 8-5).
4. Remove the foreign object with tongs.
5. Wait for the motor to cool a few minutes before you press the reset button.
6. Reconnect the electric supply, turn on the cold water and test the disposer.

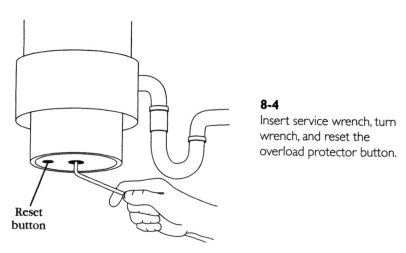

8-4
Insert service wrench, turn wrench, and reset the overload protector button.

Reset button

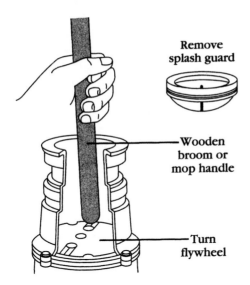

Remove
splash guard

Wooden
broom or
mop handle

8-5 Insert prying tool into the hopper.

Turn
flywheel

Installation of garbage disposer

Figure 8-6 shows some typical installations of garbage disposers. Every newly purchased garbage disposer comes with installation instructions, a use and care manual, and a warranty. Listed are the steps taken to install a new disposer.

1. Read the use and care manual.
2. Clean out the sink's drain line.
3. Disconnect electrical supply.
    a. Continuous feed disposers need a wall switch and a receptacle.
    b. Batch feed disposers need a receptacle, or must be wired directly.
4. Be sure to observe all local codes and ordinances for electrical and
    plumbing connections when installing or repairing disposers.
5. Follow the manufacturer's installation instructions.
6. After completing the installation, check for water leaks.
7. Turn on the electric, and test the operation of the garbage disposer.

GARBAGE DISPOSER MAINTENANCE

The garbage disposer is permanently lubricated; thus, it never has to be oiled. When
the disposer is used properly, it self-cleans itself. If there is an odor coming from the
inside of the disposer, you can deodorize the disposer. To deodorize the garbage
disposer, take some orange or lemon rinds and grind them up in the disposer. This
will dispel unpleasant odors, and leave the sink with a sweet smell. Another way to
deodorize the disposer is to take about a dozen or so ice cubes, sprinkled with a
generous amount of household scouring power, and grind them up in the disposer
without running the water. Flush the disposer for one minute. This will allow any
debris to be expelled into the drain.

Drain blockage

To avoid drain blockage when using the garbage disposer, allow the cold water
to flow for a sufficient time after grinding the food waste, to be sure that all of

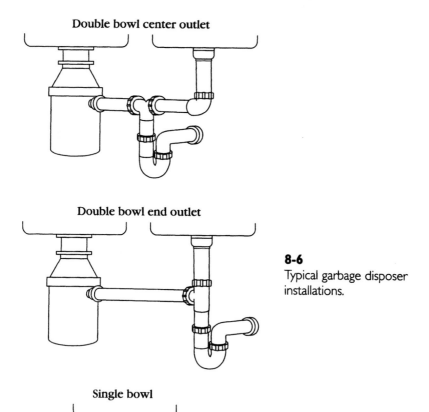

Double bowl center outlet

Double bowl end outlet

8-6
Typical garbage disposer installations.

Single bowl

the waste is flushed away. The ground waste and water mixture flows at the rate of two seconds per foot in a horizontal drain line. It is recommended that the user allows the water to flow for a minimum of 15 to 30 seconds after grinding the food waste.

The use of cold water in the garbage disposer will congeal and harden the grease, making its disposal easier. Never use chemicals, or solvent drain compounds because they can cause serious damage to the disposer.

The disposer should be used daily to flush the lines. If the dishwasher drain hose is connected to the disposer, it too should be used daily, to prevent the dishwasher drain hose from becoming clogged.

STEP-BY-STEP TROUBLESHOOTING BY SYMPTOM DIAGNOSIS

In the course of servicing an appliance, don't tend to overlook the simple things that might be causing the problem. Step-by-step troubleshooting, by symptom diagnosis, is based on diagnosing malfunctions, with their possible causes arranged into categories relating to the operation of the garbage disposer. This section is intended only to serve as a checklist to aid you in diagnosing a problem. Look at the symptom that best describes the problem you are experiencing with the garbage disposer, then proceed to correct the problem.

Overload protector trips

1. Check for voltage at the garbage disposer.
2. Does the motor hum when the disposer is turned on?
3. Was the foreign material removed from the hopper? Check and be sure that the flywheel turns easily.
4. Does the motor stay in the start winding? Use the ammeter to perform this check. Check the name plate for the correct amperage rating.
5. Did you check the overload protector?
6. Is there a shorted or open winding in the motor?
7. Is the start relay OK?

Erratic operation

1. Are there any loose connections? For example: wiring, switch, motor, etc.
2. Is the stopper worn or broken?
3. Is the wall switch defective?
4. Is the garbage disposer wired correctly?

Garbage disposer won't run

1. Did the overload protector trip?
2. Is the fuse or circuit breaker, blown or tripped?
3. Is the wall switch OK?
4. Are the motor windings burned out?
5. Check for open or shorted wiring.
6. Is the relay defective?

Garbage disposer won't stop

1. Is the wall switch defective?
2. Check for a short in the wiring.
3. Check for incorrect wiring.

Slow water flow from garbage disposer

1. Check to see if the drain is partially clogged.
2. Check the shredder teeth, and see if they are clogged with food waste. This is caused by insufficient water flow.

Garbage disposer slow grinding

1. Check for undisposable matter in the hopper.
2. Is the flywheel damaged?
3. Is there sufficient water flow?
4. Is the shredder ring worn or broken?

Abnormal noise in the garbage disposer

1. Check for undisposable matter in the hopper.
2. Are the mounting screws loose?
3. Is the flywheel broken?
4. Check the motor bearings, they might be damaged or worn.

Garbage disposer leaks

1. Disconnect the electricity. Then, locate the water leak.
2. Check for loose mounting screws.
3. Check the sink flange. There might be insufficient putty around the flange.
4. Check for leaks around the tail pipe gasket.
5. Check for holes in the hopper.
6. Is the dishwasher/disposer connector cracked and leaking?
7. Is water leaking through the motor assembly? The seals might be bad.

REPAIR PROCEDURE

The repair procedure in this chapter is a complete inspection and repair process for the garbage disposer components; containing the information you need to test a component that might be faulty; and to replace it, if necessary.

Garbage disposer assembly

The typical complaints associated with garbage disposer failure are: 1. When the motor runs, there are loud noises. 2. Water leaking from bottom of the disposer. 3. The fuses or circuit breaker will trip when you start the disposer.

1. *Verify the complaint* Verify the complaint by operating the disposer. Listen carefully, and you will hear if there are any unusual noises, or if the overload protector trips.
2. *Check for external factors* You must check for external factors not associated with the garbage disposer. Is the disposer installed properly? Does the disposer have the correct voltage?
3. *Disconnect the electricity* Before working on the garbage disposer, disconnect the electricity. This can be done by pulling the plug out of the wall receptacle. Be sure that you only remove the disposer plug. Or disconnect the electricity at the fuse panel or at the circuit breaker panel. Turn off the electricity.
4. *Remove the garbage disposer* To access to the disposer assembly, the disposer must be removed. Start with disconnecting the drain line from the disposer discharge tube. Next, remove the dishwasher drain hose

connection, if so equipped (Fig. 8-7A). Insert the service wrench into the right side of the flange body. Caution: place one hand under the disposer to keep it from falling.

Now, turn the flange to the left to free the disposer from its mounting (Fig. 8-7B). Turn the disposer upside down, and place it on a protective surface. This will protect the floor in the cabinet from damage. Remove the terminal plate (Fig. 8-7C). Remove the ground wire and wire nuts from the service cord (Fig. 8-7D, F). Separate the service cord wires from the motor wires. On models that are wired directly, loosen the cable clamp screws and remove the cable from the disposer (Fig. 8-7E).

5. *Disassemble the garbage disposer* Place the garbage disposer upside down on a protective surface, or on a workbench. Some models have an insulated outer shell that must be removed first. Most disposers will separate into four or five sections depending on the model you are servicing.

Disposers that separate into four sections consist of:
1. Top container, or hopper body
2. Upper end bell, or drain, housing assembly, cutting elements, and rotor and shaft assembly
3. Stator
4. Lower end bell assembly

Disposers that separate into five sections consist of:
1. Top container or hopper assembly
2. Stationary shredder
3. Upper end bell, or drain, housing assembly, rotor shredder, and rotor and shaft assembly
4. Stator
5. Lower end bell assembly

Before removing any screws from the disposer, you must place scribe marks on the lower end bell and stator, the upper end bell and stator, and the upper end bell and hopper. These scribe marks will be used to align the sections properly when the garbage disposer is reassembled. Remove the four through-bolts from the lower end bell assembly and separate the sections of the disposer. Label the wires from the stator to the lower end bell assembly. Disconnect them. Be cautious if the garbage disposer has a capacitor.

Warning: do not touch capacitor terminals. A charged capacitor can cause severe shock, when both terminals are touched. A charged capacitor will hold a charge until it is discharged.

6. *Inspect and test the garbage disposer components* Inspect the hopper assembly for cracks, holes, and (on some models) corrosion around the dishwasher connection. Also, inspect the gasket area for nicks or rough spots. Inspect the gasket or "O" ring for cuts, breaks, or worn areas. Replace it if it is defective. Inspect the upper end bell, or drain, housing assembly for the same. Test the centrifugal switch for binding by manually operating it. It should slide up and down freely. Inspect the rotor and thrust washer for cracks, breaks, or wear. If the upper drain housing and rotor assembly failed inspection, replace the garbage disposer as a complete unit.

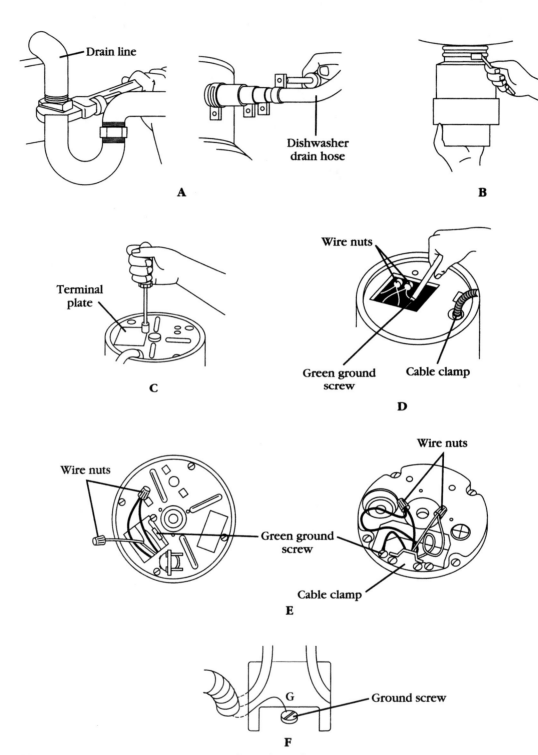

Drain line

Dishwasher
drain hose

A

B

Terminal
plate

C

Wire nuts

Green ground
screw

Cable clamp

D

Wire nuts

Wire nuts

Green ground
screw

Cable clamp

E

G

Ground screw

F

8-7 Illustrating the steps taken to remove the garbage disposer.

Before testing the capacitor, it must be discharged. Use a screwdriver with an insulated handle. Discharge the capacitor by shorting across both terminals. Remove the wire leads, one at a time, with a pair of needle nose pliers. Set the ohmmeter on the highest scale, place one probe on one terminal, then attach the other probe to the other terminal (Fig. 8-8). Observe the meter action. While the capacitor is charging, the ohmmeter will dip toward zero ohms, for a short period of time. Then, the ohmmeter reading, will return to infinity. If the ohmmeter reading deflects to zero, and does not return to infinity, the capacitor is shorted, and should be replaced. If the ohmmeter reading remains at infinity, the capacitor is open, and should be replaced.

Inspect the stator windings for nicks, cuts, or burned spots. Set the ohmmeter on R × 1, and test the stator windings for open, grounded, or shorted windings. Test both the start and run windings. If the stator windings fail the test, replace the garbage disposer as a complete unit. Set the ohmmeter on R × 1, and test the start switch. Place the probes on the wires leading to the switch. To ensure an accurate meter reading, isolate the start switch wires from the remainder of the circuitry. The meter reading will indicate an open circuit. If the test fails, replace the start switch. Next, test the overload protector. Isolate the overload protector wires from the remainder of the circuitry. Then, place the ohmmeter probes on the overload protector wires. Push the red reset button in on the overload protector. The ohmmeter reading will indicate a closed circuit. If this test fails, replace the overload protector.

7. *Reassemble and reinstall the garbage disposer* To reassemble the disposer, just reverse the disassembly procedure, and reassemble. Check to be sure

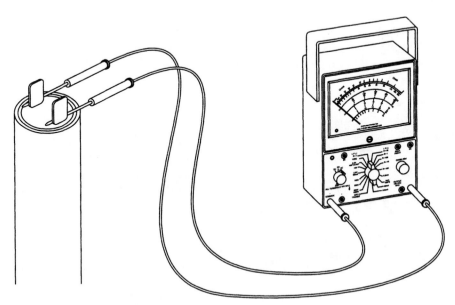

8-8 Place ohmmeter test leads on the capacitor terminals.

that all wiring is correct. Remember to use the scribe marks to align the sections properly before inserting the through-bolts. Also, be careful not to damage any of the gaskets or "O" rings when assembling the garbage disposer. Reconnect the service cord wires to the motor wires, and tighten the cable clamp screws (Fig. 8-7E). Next, reconnect the ground wire to the green ground screw (Fig. 8-7D, F). Reinstall the terminal plate and screw (Fig. 8-9A).

Lift the disposer, and position it so that the disposer's three mounting ears are lined up under the ends of the sink mounting assembly screws (Fig. 8-9B, C). The disposer will now hang by itself. After the plumbing is

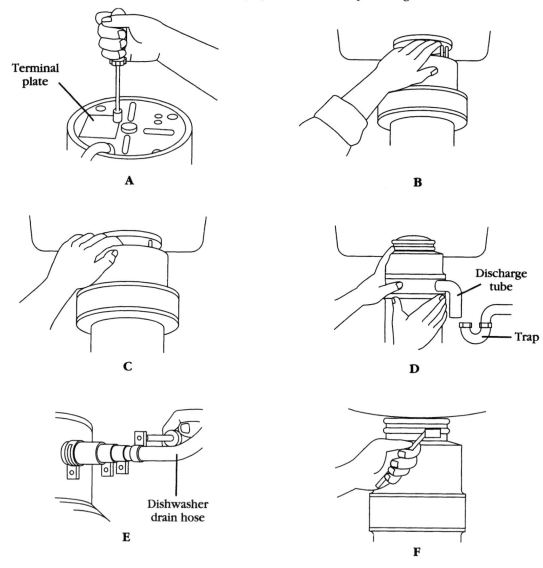

8-9 Illustrating the steps taken to reinstall the garbage disposer.

reconnected, you will then tighten the ring using the service wrench. Rotate the disposer to align the discharge tube with the drain trap (Fig. 8-9D). Reconnect them. Next, reconnect the dishwasher drain hose connection, if so equipped (Fig. 8-9E). Finally, place the service wrench into the left side of one of the disposer mounting lugs, located on the top of the disposer. Then, turn the wrench to the right until it is firmly secured, engaging the locking notch (Fig. 8-9F).

Before you reconnect the electricity, check for leaks by running the water. Inspect all connections. Inspect the disposer for water leaks also. If no leaks are present, turn on the electric, and run the disposer; reinspect for any water leaks.

Before installing a new garbage disposer, the dishwasher connection plug must be removed, if the defective disposer was so equipped with a dishwasher drain hose connection. The dishwasher connection plug is removed by knocking the plug out with a hammer and some blunt instrument, such as a dowel or a punch (Fig. 8-10). Do not use a screwdriver or any sharp tool. This can be driven through the plug, and make it difficult to remove. Remove the plug from the hopper before installing or operating the disposer.

DIAGNOSTIC CHARTS

The following diagnostic flow charts and wiring diagrams will help you to pinpoint the likely cause of the problem (Figs. 8-11, 8-12, 8-13).

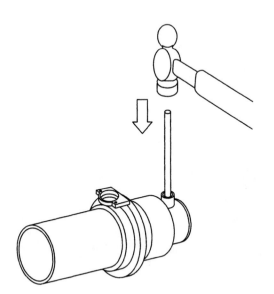

8-10
Before installing a new garbage disposer, remove the dishwasher connection plug if a dishwasher drain hose will be connected to the disposer.

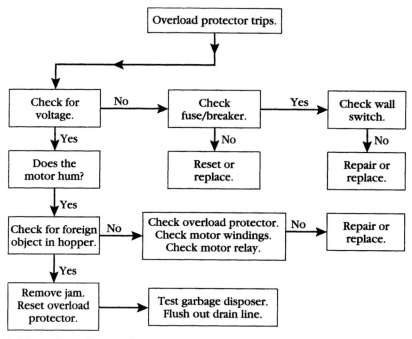

8-11 Garbage disposer flow chart: overload protector trips.

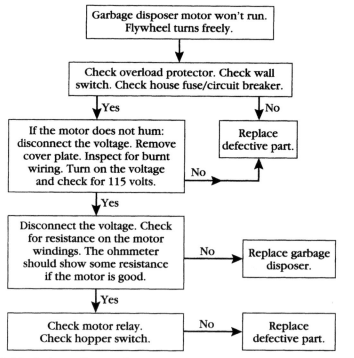

8-12 Garbage disposer flow chart: garbage disposer motor won't run.

Typical household garbage disposer wiring diagrams

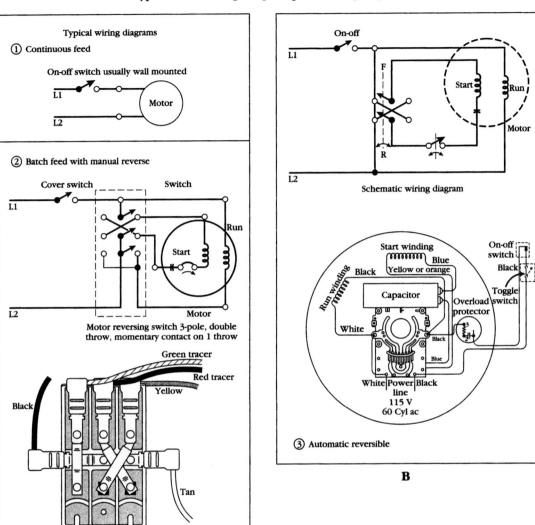

8-13 Typical household garbage disposer wiring diagram.

9

Electric
water heaters

Electric water heaters are heat-producing appliances. They normally have two heating elements (Fig. 9-1), with each element controlled by a thermostat. The thermostats are mounted on the outer wall of the water heater tank just above the elements, from where they sense the temperature of the water through the outer wall of the water heater tank. A temperature and pressure relief valve, which is mounted on the outside of the tank, is a device applied to a water heater which will open to pass water or steam, if excessive pressure or temperature occurs in the water heater tank.

Electric water heaters are available in many different sizes of height and width. They are similarly available in different capacities.

PRINCIPLES OF OPERATION

The water heater tank is full of water; the electricity is on; the upper thermostat (Fig. 9-1) senses that the water is cold; and this condition energizes the upper element. At the same time, the lower element is not activated. The upper heating element will heat approximately one quarter of the tank's capacity. When the temperature of the water reaches the thermostat setting, the upper thermostat will shut off. Then, the lower thermostat becomes energized and heats the remainder of the water in the tank. When the temperature of the water reaches the thermostat setting, the lower thermostat will deactivate the lower heating element. The tank is now filled with hot water. As the consumer uses the hot water, hot water is drawn from the top of the tank, and is replaced with cold water through a dip tube (Fig. 9-1) located near the bottom of the tank. When the lower thermostat senses the cold water, the thermostat activates the lower heating element, heating the water. Figure 9-2 identifies one of the many types of thermostats. Also, in the same figure, there is a schematic drawing showing the thermostat switch contacts.

SAFETY FIRST

Any person who cannot use basic tools, or follow written instructions, should *not* attempt to maintain or repair any electric water heaters. Any improper installation, preventive maintenance, or repairs could create a risk of personal injury or property damage.

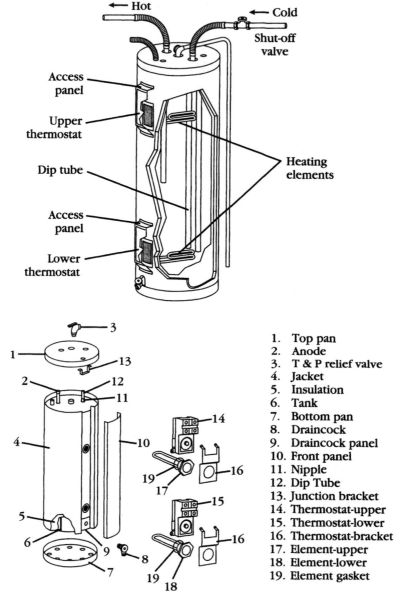

← Hot

← Cold

Shut-off
valve

Access
panel

Upper
thermostat

Dip tube

Access
panel

Lower
thermostat

Heating
elements

1. Top pan
2. Anode
3. T & P relief valve
4. Jacket
5. Insulation
6. Tank
7. Bottom pan
8. Draincock
9. Draincock panel
10. Front panel
11. Nipple
12. Dip Tube
13. Junction bracket
14. Thermostat-upper
15. Thermostat-lower
16. Thermostat-bracket
17. Element-upper
18. Element-lower
19. Element gasket

9-1 Exploded view of the electric water heater.

If you do not fully understand the preventive maintenance or repair procedures in this chapter, or if you doubt your ability to complete the task on your electric water heater, then please call your service manager.

These precautions should also be followed:
1. Before checking the water heater, turn off the electricity.
2. Never restore the electricity to the water heater if the tank is empty; do so only after it is full of water.

3. Never remove the heating elements with the tank full of water.
4. If a water heater needs to be replaced, it is strongly recommended that all electrical, plumbing, and placement of the tank should be done by qualified personnel. Observe all local codes and ordinances for electrical, plumbing, and installation procedures.

Before continuing, take a moment to refresh your memory concerning the safety procedures in Chapter 2.

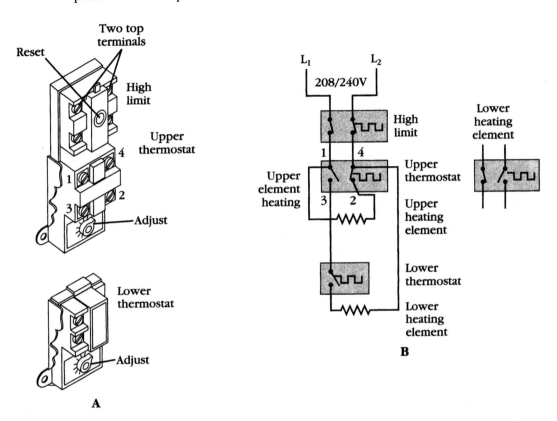

9-2 A. Typical upper and lower water heater thermostats. B. Wiring schematic for upper and lower thermostat.

ELECTRIC WATER HEATERS IN GENERAL

Much of the troubleshooting information in this chapter covers electric water heaters in general, rather than specific models, in order to present a broad overview of service techniques. The pictures and illustrations that are used in this chapter are for demonstration purposes; for clarifying the description of how to service these heaters. In no way do they reflect on a particular brand's reliability.

ELECTRIC WATER HEATER MAINTENANCE

Every so often, inspect the water heater tank for possible water leaks. Check the following:

1. Check all pipe connections to the tank. If corroded, they must be repaired before there is serious water damage to the property.
2. Check the temperature/pressure relief valve. Is it leaking?
3. Turn off the electricity, remove the access panels, inspect the wiring, elements, and insulation for signs of water leakage. If water is leaking, the leak must be repaired, or the unit must be replaced immediately.

Once a year the water heater tank should be drained and flushed out. By doing this procedure, it will increase the life expectancy of the tank by removing the unwanted sediment. Before you begin draining the tank, turn off the electricity to the water heater. To drain the water heater tank, turn off the water supply to the tank. Connect a garden hose to the draincock, and open the valve. To increase the water flow draining out of the tank, open the hot water faucet, this will increase the flow of water draining out of the tank. When you are ready to refill the tank, close the draincock, turn on the water supply to the water heater, and begin to refill the tank. Leave the closest hot water faucet open during refilling; the air that is trapped in the tank will escape through the hot water faucet. When all of the air is out of the tank and water lines, close the hot water faucet. Go to all of the other hot water faucets and open them to remove the air in the lines. Close all water faucets. Now you are ready to turn on the electricity to the water heater.

STEP-BY-STEP TROUBLESHOOTING BY SYMPTOM DIAGNOSIS

In the course of servicing an appliance, don't tend to overlook the simple things which may be causing the problem. Step-by-step troubleshooting, by symptom diagnosis, is based upon diagnosing malfunctions with their possible causes arranged into categories relating to the operation of the water heater. This section is intended only to serve as a checklist to aid you in diagnosing a problem. Look at the symptom that best describes the problem that you are experiencing with the water heater, then proceed to correct the problem.

No hot water

1. Do you have the correct voltage at the water heater?
2. Check for loose wiring.
3. Check to be sure the reset button did not trip.
4. With the electricity off, check the thermostats for continuity.
5. With the electricity off, check the heating elements for continuity. Also check for a grounded heating element.

Not enough hot water

1. Is the water heater undersized for the usage?
2. Check the lower heating element for continuity. Also check for a grounded heating element.
3. Check for leaking faucets and pipes.
4. Check to see if anyone is wasting water.
5. Drain the tank; remove lower element; see if any sediment is in the bottom of tank, or around the heating element.

6. Is there any lime formation on the elements?
7. Are the thermostats operating properly? Run a cycle test.
8. Check the thermostat settings. Verify if they are set too low.

Water too hot

1. Check to see if the thermostat is snug against the tank.
2. Is the thermostat set too high?
3. Are the thermostats operating properly? Run the cycle test.
4. Check for a grounded element.

Water heater element failure

1. Check for loose or burned wiring connections.
2. Check for the correct voltage to the elements.
3. Is the element shorted or grounded? Test the elements for continuity.

Discolored or rusty water

1. Drain the tank; remove the lower element; see if there is any sediment in the bottom of the tank.
2. Is the water supply to the water heater rusty in color?
3. Check the water for softness.
4. Check to see if there are excessive mineral deposits.
5. Inspect the anode rod to note its condition. If it did dissipate, replace the anode.

Repair procedures

Each repair procedure is a complete inspection and repair process for a single water heater component; containing the information you need to test a component that may be faulty; and to replace it, if necessary.

Upper thermostat

The typical complaints associated with the upper thermostat are: 1. No hot water. 2. Burned wires. 3. Water heater runs continuously. 4. The fuse blows, or the circuit breaker trips.

1. *Verify the complaint* To verify the complaint, turn on the hot water. Does the water get hot?
2. *Check for external factors* You must check for external factors not associated with the water heater. Is the water heater installed properly? Does the water heater have the correct voltage?
3. *Disconnect the electricity* Before working on the water heater, disconnect the electricity. This can be done by pulling the plug out of the electrical outlet. Or disconnect the electricity at the fuse panel or at the circuit breaker panel. Many installations have a disconnect switch/breaker box near the heater. Turn off the electricity.

4. *Remove the access panel* To gain access to the upper thermostat, the access panel must be removed (Fig. 9-1). The access panel is held on with 2 screws on most models. Peel back the insulation; then, remove the thermostat protective cover.

5. *Test the thermostat* To test the upper thermostat, use your voltmeter. Set the range on the 300-volt scale. Touch the probes to terminals 1 and 3 above the reset button (Fig. 9-3). Turn on the electricity. If the meter reads no voltage, have the customer call an electrician to find out where the power failure is. If the meter reads the proper rated voltage, as stated on the name plate, the electricity is okay. Next, touch the probes to terminals 2 and 4 below the reset button (Fig. 9-4). If the meter reads the proper rated voltage, the electricity is okay. If the meter reading is "0" volts, press the reset button on the thermostat. If there is still no voltage reading on the meter scale, either the water temperature in the tank is too hot, or the thermostat is inoperative. At this stage, again turn off the electricity to the tank. Remove the wires from the upper heating element (Fig. 9-5). Be absolutely sure that the electricity is OFF before any further tests are made using your ohmmeter.

Use your ohmmeter to test terminals 1 and 2 on the upper thermostat (Fig. 9-5). If the thermostat is calling for heat, the ohmmeter needle will swing to the right showing continuity. Now, touch the probes to terminals 1 and 4 on the thermostat (Fig. 9-6). If the water in the tank is cold, you will not get a reading on the ohmmeter scale. If the water in the tank is hot, as called for by the thermostat, the needle will swing to the right, thus showing continuity.

6. *Cycle the upper thermostat* Reconnect the wires to the heating element. Turn on the electricity to the tank. Take your ammeter and clamp the jaws around the wire that goes from terminal number 4 to the upper element

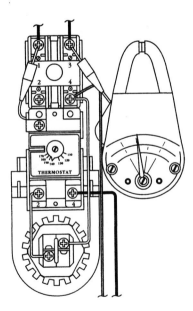

9-3
Place the voltmeter probes on terminals 1 and 3 to check the voltage.

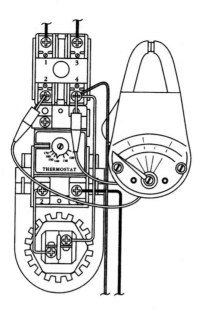

9-4
Place the voltmeter probes on terminals 2 and 4 to check the voltage.

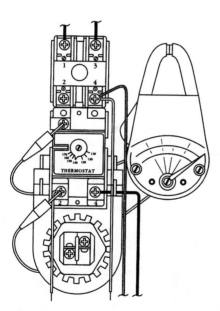

9-5
Remove the wires from the element. Place the ohmmeter probes on terminals 1 and 2. When the thermostat is calling for the element to turn on, the ohmmeter scale will read continuity.

(Fig. 9-7). With cold water in the tank, and the thermostat calling for heat, you should get a reading on the meter in amps. The reading you get will depend on the rating (in watts) of the heating element, divided by the voltage supplied (see the formula in Fig. 9-8). After the temperature is reached, the needle on the ammeter scale will read "0."

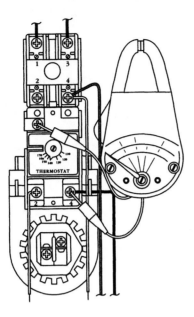

9-6
Place the ohmmeter probes on terminals 1 and 4. If the water in the tank is cold, the ohmmeter scale will read no continuity.

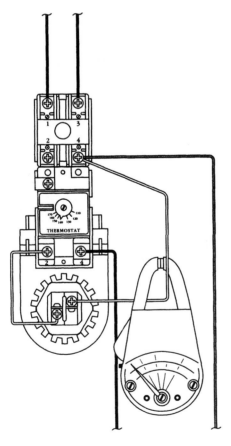

9-7
The ammeter jaws encircled on the number 4 terminal wire go to the element.

Conversion chart for determining amperes, ohms, volts, or watts
(Amperes = A, Ohms = Ω, Volts = V, Watts = W

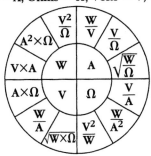

9-8 Ohm's Law equation wheel: the conversion chart for determining amperes, ohms, volts, or watts.

Next, place the ammeter jaws on the number 4 terminal wire, that goes from the upper thermostat (just below the temperature gauge setting) to the lower element (Fig. 9-9). The ammeter should show a reading, indicating that the lower element is working.

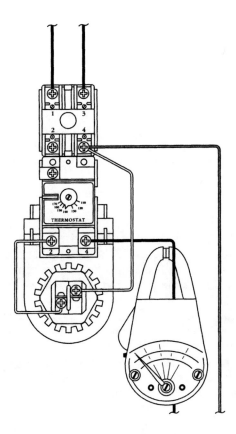

9-9
Place the ammeter jaws around terminal 4. The ammeter reading is 0.

7. *Replace the upper thermostat* Turn off the electricity to the water heater. To replace the upper thermostat, remove all of the wires from the thermostat. Be sure that you mark the wires so that you can replace them exactly the way you took them off. Next, pry back the thermostat bracket with your fingers far enough to slide the thermostat up and out; remove the thermostat. To install the new thermostat, just reverse the disassembly procedure, and reassemble. Be sure that the thermostat is snugly mounted against the tank so that it can sense the temperature better.

8. *Test the new upper thermostat* After you have completed the new installation, turn on the electricity to the water heater. Repeat this testing process as described, and verify that the new unit is working correctly.

Lower thermostat

Testing the lower thermostat is similar to testing the upper thermostat; with the following exception. With the wires removed from the element, place the ohmmeter's probes on each terminal. Take a screwdriver and place it on the set screw of the thermostat, and turn it clockwise. The needle on your ohmmeter will swing to the right, indicating that there is continuity. If the needle does not move, the thermostat needs to be replaced. Replace this thermostat as per the upper thermostat replacement previously described. Then, test the thermostat by placing the jaws of the ammeter around the wire that connects to the lower thermostat. Check to be sure that the thermostat cycles on and off.

Heating element

The typical complaints associated with the heating element are: 1. No hot water. 2. Burned wires. 3. The fuse blows, or the circuit breaker trips.

1. *Verify the complaint* To verify the complaint, turn on the hot water. Does the water get hot?

2. *Check for external factors* You must check for external factors not associated with the water heater. Is the water heater installed properly? Does the water heater have the correct voltage?

3. *Disconnect the electricity* Before working on the water heater, disconnect the electricity. This can be done by pulling the plug out of the electrical outlet. Or disconnect the electricity at the fuse panel, the circuit breaker panel, or the disconnect switch. Turn off the electricity.

4. *Remove the access panel* In order to gain access to the heating element, the access panel must be removed (Fig. 9-1). The access panel is held on with two screws on most models. Peel back the insulation; then, remove the wires from the heating element.

5. *Test the heating element* Use the ohmmeter to test the heating element. Set the range to R × 1. Touch the probes to the element screws (Fig. 9-10). If the element is good, the ohmmeter scale will show continuity. If the element is bad, the ohmmeter needle will not move, showing an open element. Now, set the ohmmeter on R × 100. Take one probe and place it on either terminal of the element. Take the other probe and touch the element head (Fig. 9-11). If you get any reading, the element is grounded,

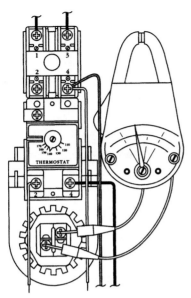

9-10 Testing heater element. Place the ohmmeter probes on the screw terminals of the element.

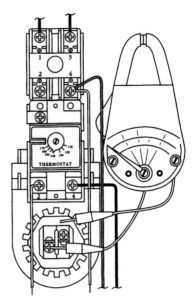

9-11 Testing the heater element for the ground.

and it should be replaced. Repeat this procedure for the other terminal on the element. Both terminals must be measured for grounding.

6. *Remove the element* Turn off the electricity to the water heater. Next, turn off the water supply. Connect a garden hose to the draincock, and open the draincock. At the same time, open a nearby faucet to allow the water to drain faster. After the tank is empty, disconnect the wires from the element and remove it.

7. *Install the new element* To install the new element, just reverse the disassembly procedure, and reassemble. When installing the new element, always replace the gasket. Before you insert the new element into the tank, clean the flange of debris.

8. *Test the new element* When you are done installing the replacement element, close the draincock, and open the water supply. Close the nearby faucet after all of the air in the tank has dissipated. Next, check for water leaks. Turn on the electricity to the water heater. Use the ammeter to check the element. The ammeter should show a reading.

DIAGNOSTIC CHARTS

The following diagnostic flow charts, wiring diagrams, and tables will help you to pin-point the likely causes of the problem (Figs. 9-12 through 9-19; Tables 9-1 and 9-2).

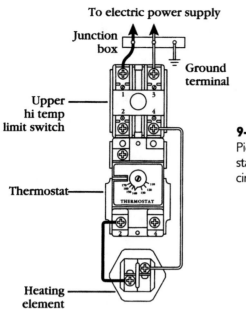

To electric power supply

Junction box

Ground terminal

Upper hi temp limit switch

Thermostat

Heating element

9-12
Pictorial wiring diagram for a standard 2-wire 208-/240-volt circuit, single element.

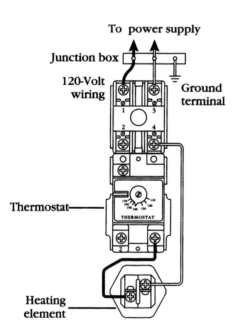

To power supply

Junction box

120-Volt wiring

Ground terminal

Thermostat

Heating element

9-13
Pictorial wiring diagram for a standard 2-wire 208-/240-volt circuit, single element.

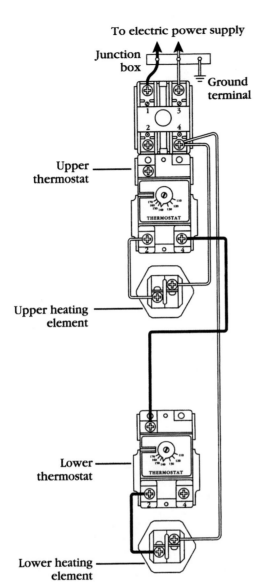

To electric power supply

Junction box

Ground terminal

1   3
2   4

Upper thermostat

170 110
160 120
150 140 130
THERMOSTAT

2   4

Upper heating element

9-14
Pictorial wiring diagram for a standard 2-wire interlocking 208-/240-/277-volt circuit, double element.

Lower thermostat

170 110
160 120
150 140 130
THERMOSTAT

2   4

Lower heating element

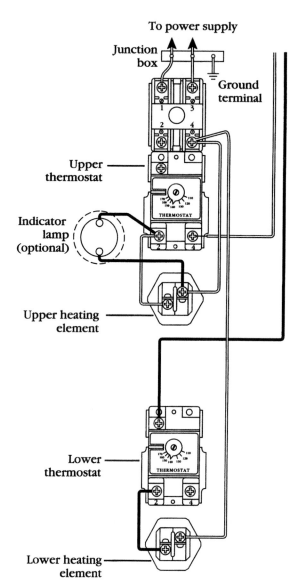

To power supply

Junction
box

Ground
terminal

1   3

2   4

Upper
thermostat

THERMOSTAT

170   110
160   120
150 140 130

Indicator
lamp
(optional)

2   4

Upper heating
element

9-15
Pictorial wiring diagram for a
4-wire off-peak double-element
nonsimultaneous operation.

Lower
thermostat

170   110
160   120
150 140 130

THERMOSTAT

2   4

Lower heating
element

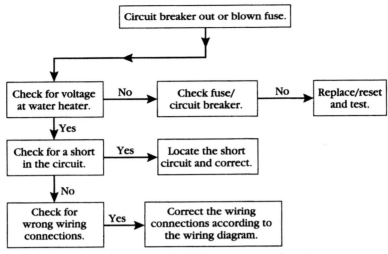

9-16 Water heater flow chart: circuit breaker out or fuse blown.

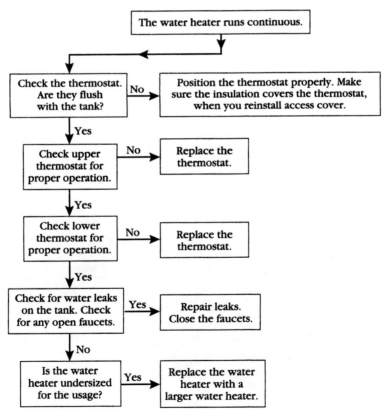

9-17 Water heater flow chart: water heater runs continuously.

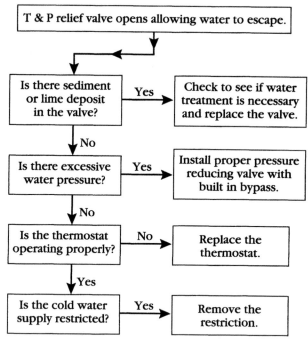

9-18 Water heater flow chart: relief valve opens.

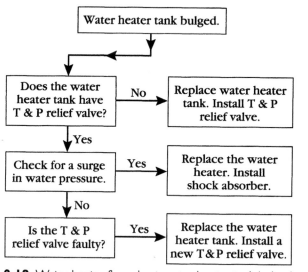

9-19 Water heater flow chart: water heater tank bulged.

Table 9-1

Typical Hotwater Usage Chart

Activity	Gallons
Handwashing	0.9
Shower	5–7
Tub bath	7–12
Shaving	2
One meal per person	2–5
Automatic washer	11–34
Conventional washing	30
Dishwasher	8
Average bathroom use per day	
Adult	12
Child	24

Table 9-2

Electric Water Heater Recovery
(Calculated at 100% recovery efficiency)

GPH recovery at indicated temperature rise					
Heating element wattage	degrees in Fahrenheit				
	60	70	80	90	100
750	5.1	4.4	3.8	3.4	3.1
1000	6.8	5.8	5.1	4.5	4.1
1250	8.5	7.3	6.4	5.7	5.1
1500	10.2	8.8	7.7	6.8	6.1
2000	13.7	11.7	10.2	9.1	8.2
2500	17.1	14.6	12.8	11.4	10.2
3000	20.5	17.5	15.4	13.6	12.3
3500	23.9	20.5	17.9	15.9	14.3
4000	27.3	23.4	20.5	18.2	16.4
4500	30.7	26.3	23.0	20.5	18.4
5000	34.1	29.2	25.6	22.7	20.5
5500	37.6	32.2	28.2	25.0	22.5
6000	41.0	35.1	30.7	27.3	24.6

10
Automatic washers

The automatic washer is a complex electromechanical machine (Fig. 10-1). It performs various cycles to clean clothes. There are times when a washer fails to operate properly. Don't let its complexity intimidate you. This chapter will provide the basics needed to diagnose and repair the washing machine.

PRINCIPLES OF OPERATION

The clothes are placed evenly into the washer basket, being sure that the washer is not overloaded, and that the proper cycle is selected.

The user activates the washer through the timer. The internal switches of the timer distribute the electricity to activate the other components in the washing machine during a given time period, designated by the internal cam of the timer.

The water enters the tub through the water fill hoses, the water inlet valve, and water inlet hose. Hot, warm, or cold water is selected by the user via the water temperature selector switch, located on the console panel. On some models, the water temperature selection is controlled by the timer.

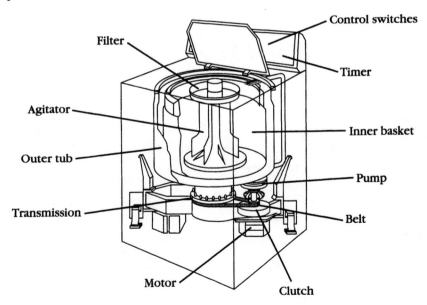

10-1 Parts identification and location in an automatic washer.

The amount of water that fills the tub, is controlled by the water level control (pressure switch). The water level control offers a choice of water levels depending on the amount of clothing being washed. As the water level rises in the tub, it forces air through the air dome and up the plastic tube to the water level control. The pressure that is exerted on the water level control's diaphragm, will trip the water level switch from empty to full, supplying electricity to the washer drive motor and, thus, operating the transmission.

The transmission is operated by the drive motor, either by belt drive or by direct drive, depending upon the model. Agitation is accomplished by the agitator, which is located in the center of the basket, and which is driven by the transmission. The agitator turns clockwise and counterclockwise, creating a water motion that moves the clothes within the basket.

When the washer goes into the drain mode, the agitator will stop agitating, and the water leaves the bottom of the tub through the water pump, to be pumped into the drain. The water pump may operate by belt drive, or by direct motor drive.

A final deep rinse cycle (the tub fills to the selected water level, and begins to agitate) will be introduced, to wash off any remaining residue of soap or dirt.

The timer will now select the spin cycle, and the washer will then go into the spin mode. In the spin mode, the washer spins the clothes, to remove most of the water out of the clothing by centrifugal force. Some models use brief sprays of water to remove any residue of soap or dirt remaining on the clothes in the spin cycle.

Functions and cycles

The removal of soil from clothing and fabrics is accomplished by a combination of mechanical and chemical processes.

1. *Mechanical process* Soil is removed by agitating and by forcing the detergent through the clothing.
2. *Chemical process* The detergent used will dissolve and loosen the soil in the clothing. As the washing machine operates through its cycles, it is aided by hot, soft water, which increases the chemical processes of the detergent being used.

Washing machines perform four basic functions, that are modified and put together in different ways, to create the various cycles. The four functions are:

1. Fill
2. Agitate
3. Drain
4. Spin

SAFETY FIRST

Any person who cannot use basic tools, or follow written instructions, should *not* attempt to install, maintain, or repair any automatic washers. Any improper installation, preventive maintenance, or repairs could create a risk of personal injury or property damage.

If you do not fully understand the installation, preventive maintenance, or repair procedures in this chapter; or if you doubt your ability to complete the task on the automatic washer, then please call your service manager.

The following precautions should also be followed:
1. Never bypass, or disconnect, any part or device (originally designed into the washer) as a temporary repair.
2. Always reconnect all ground wires, and be sure that they are secure.
3. Be careful of moving parts and sharp edges.

Before continuing on, take a moment to refresh your memory of the safety procedures in Chapter 2.

AUTOMATIC WASHERS IN GENERAL

Much of the troubleshooting information in this chapter covers automatic washers in general, rather than specific models, in order to present a broad overview of service techniques. The pictures and illustrations that are used in this chapter are for demonstration purposes; to clarify the description of how to service washing machines, in no way to reflect on a particular brand's reliability.

Location and installation of automatic washer

Listed are some general principles that should be followed when performing the installation of a washing machine.
1. Locate the washing machine where there is easy access to existing drain, water, and electrical lines.
2. Be sure you observe all local codes and ordinances for the electrical and plumbing connections.
3. The washing machine should be installed and leveled on a firm floor to minimize vibration during operation.
4. Do not install the washing machine in an area where the temperature might be below freezing.
5. To reduce the risk of a fire, never install a washing machine on any type of carpet.
6. Always follow the installation instructions that are provided with every new washing machine model purchased.

Water supply

The water supply for an automatic washer should have a hot and cold faucet, located within 5 to 7 feet of the washer. The faucets should be ¾"-threaded type to accept the fill hose connection.

The water pressure must be between 25 and 125 pounds per square inch for the washer to operate properly. The water coming out of the fill hoses should be equal in both pressure, and in the volume of water, to prevent unacceptable water temperature changes when entering and filling the washer.

The hot water supply to the washer should be between 130 and 150 degrees Fahrenheit. If the hot water temperature is below 70 degrees Fahrenheit, the clothes being washed will not clean properly, and the detergent will not dissolve properly. You can check the temperature of the hot water by operating the washer in the fill mode, with the water temperature setting on hot. Let the water run until it is as hot as possible, then insert a thermometer into the stream of water. If the thermometer reading is below 130 degrees Fahrenheit, then you will have to raise the water heater thermostat setting.

The cold water temperature should be between 70 and 100 degrees Fahrenheit. When the washer is in its rinse stage, the cold water will prevent wrinkles from setting into the fabrics. Some fabric manufacturers require that their fabrics be washed in cold water, both to prevent shrinkage, and to eliminate the possibility of destroying the fabric. When the user selects the warm fill, the temperature of the water should be 100 degrees Fahrenheit.

It is recommended that the consumer read the use and care manual before performing a wash. Most use and care manuals have a water temperature guide to assist the user in the proper selection of the water temperature.

Drain requirements

The drain, to which the washer's drain hose is connected, must be able to accept at least a 20 to 30 gallons per minute flow, in order to remove the water out of the tub. The standpipe should be at least 32 inches high, and not to exceed 60 inches in height. The internal diameter of the drain pipe should be a minimum of 1½ inches, in order to provide an air gap around the drain hose and, thus, to prevent the suction from siphoning the water out of the tub during the following cycle.

Detergent

The kind, and the amount, of detergent that is used is an important part of getting the clothes clean. Different brands of detergent contain different amounts of phosphorous, which works to soften the water and to boost the cleaning action. If the water is hard, you might need to recommend a detergent with a higher phosphorous content. If the water is soft, the user can use a low-phosphorous detergent. Some areas restrict the phosphate content to 8%, or even less. This means that the user will have to increase the amount of detergent used, in those areas where the water is hard. This is done by adding a certain extra amount of detergent manually, to the wash cycle.

It is recommended that the consumer read the use and care manual before performing a wash. Most use and care manuals have a detergent guide, to assist the user in the recommended amount of detergent to be used.

Water hardness is measured in grains.
- 0 to 3 grains, soft water
- 4 to 9 grains, medium hard water
- 10 to 15 grains, hard water
- over 15 grains, very hard water. If you do not know the hardness of the water supply, contact the local water department.

STEP-BY-STEP TROUBLESHOOTING BY SYMPTOM DIAGNOSIS

In the course of servicing an appliance, don't overlook the simple things that might be causing the problem. Step-by-step troubleshooting, by symptom diagnosis, is based on diagnosing malfunctions with their possible causes arranged into categories relating to the operation of the washer. This section is intended only to serve as a checklist to aid you in diagnosing a problem. Look at the symptom that best describes the problem that you are experiencing with the washer, then proceed to correct the problem.

No water entering washer

1. Is the washer plugged in?
2. Check for proper voltage at the wall receptacle.
3. Is the water supply turned on? The fill hoses should feel stiff.
4. Test the water temperature switch contacts for continuity.
5. Check for an open circuit in the timer contacts.
6. Check for loose wires to the water valve solenoid.
7. Test water valve solenoid coils for continuity.
8. Check the water valve inlet strainer screens. Remove the fill hoses to inspect these filters.
9. Test the water level control switch for continuity.
10. Are the water supply hoses kinked?

Motor will not run

1. Check for a proper voltage at the wall receptacle.
2. Is the washer plugged in?
3. Check for a blown fuse or a tripped circuit breaker.
4. Check for a faulty timer.
5. Are there any loose wires to the timer, motor, etc.?
6. Test the motor windings for continuity.
7. Test the thermal overload in the motor for continuity.
8. Test the water level control switch contacts for continuity.
9. Test the motor speed selector switch contacts for continuity.
10. Are there any open wires in the washer circuit?
11. Test the capacitor on the motor.
12. Check the centrifugal switch in the motor.
13. Check for continuity of the lid switch contacts.

Washer will not agitate

1. Check for a broken or worn belt.
2. Check the motor to transmission drive coupling.
3. Test the timer contacts for continuity.
4. Are there any loose wires within the wiring harness?
5. Are there any loose wires to the timer, motor, etc.?
6. Test the continuity of the motor windings.
7. Test the agitator solenoid on the transmission.
8. Check for loose pulleys on the transmission and motor.
9. Check the water level control switch.
10. Test for continuity of the start relay.
11. Test the capacitor on the motor.

12. Check the centrifugal switch in the motor.
13. Test for continuity of the lid switch contacts.
14. Test for continuity of the speed selector switch.
15. Check the clutch assembly.
16. Check the transmission.

Water will not drain

1. Check for a clogged drain connection.
2. Inspect the pump for obstructions.
3. Check the drain hose, and be sure it is not kinked.
4. Check the belt that goes to the pump.
5. Check for suds lock. If this happens, just add cold water and flush the suds out of the pump. (Suds lock is caused by too much soap remaining in the tub, pump, and the connecting hoses. This condition will prevent water from draining effectively.)
6. Check the pump coupling. The pump and motor must be removed for a visual inspection of the coupling.
7. Check for air lock in the pump (air trapped inside the pump, caused by debris).
8. Check to be sure that the motor is not running in the agitation direction.
9. Check the height of the drain.
10. Does the pump pulley turn freely?

Washer will not spin

1. Check for a loose or broken belt.
2. Check for loose pulleys.
3. Check the clutch assembly.
4. Check for loose or broken wires in the washer circuit.
5. Test the continuity for a faulty lid switch assembly.
6. Test the continuity for a defective spin solenoid.
7. Check for a broken drive coupling. The pump and motor must be removed for a visual inspection of the coupling.
8. Test the continuity of the water level control switch.
9. Test the continuity of the speed selector switch.
10. Test the continuity in the motor windings and the motor overload protector.
11. Test the timer contacts for continuity.
12. Check for clothing jammed between the inner basket and the outer tub.
13. Check all seals and mechanical linkages.
14. Check the transmission.
15. Check to be sure that the motor is running in the spin direction.

COMMON WASHABILITY PROBLEMS

If there are no mechanical problems with the washer's operation, and the complaints are that the washer does not clean the clothes properly, you have a washability problem. The next step should be to look at the cause that best describes the problem that the customer is experiencing with the washer. Then, proceed to correct the problem. If necessary, instruct the user how to get better results from their automatic washer.

Stains on the clothing

Stains on clothing can be caused by a number of different things. As the servicer, you will have to determine if it is caused by the washer's components or from an external source. Many stains are blamed on leaking transmissions. This type of problem is related to the increasing use of synthetic fabrics, and to the poor washing practices of the user. Many of these stains consist of cooking oil or grease, and are not visible when they first occur during cooking or eating. The oil that is embedded in the clothing acts like glue, attracting dirt from the wash water. When the wash cycle is completed, the clothes come out dirty and spotted. If the transmission oil leaked into the wash water, there would be stains on all of the clothes in a random pattern. The color of transmission oil embedded into the clothing is usually a brownish-yellow stain. Transmission oil cannot be washed out of the clothes, it requires a solvent to remove the stains.

Listed below are some stain removal rules for clothing in general:

1. Stains are easier to remove when they first appear on the clothing. If the stains are old, they might never come out of the clothing.
2. Before attempting to remove any stain, you must know:
   a. what type of stain,
   b. what kind of fabric, and
   c. how old is the stain?
3. Use only cold or warm water to remove stains. Hot water will set the stain permanently into the fabric.
4. When bleach is recommended for the removal of the stain, use a bleach that is safe for the fabric. When using a chlorine bleach, always dilute it with water to protect the bleach from destroying the fibers.
5. Always test stain remover products on a hidden corner of the garment to see if the color remains in the fabric.
6. When preparing to remove the stain from the fabric, face the stained area down on a paper towel or a white cloth. Then, apply the stain remover to the back of the stain so that the stain will be forced off the fabric, instead of through the fabric.
7. Some protein stains can be removed with an enzyme pre-soak, or with meat tenderizer.
8. When using dry cleaning solvents, always use them in a well-ventilated room, away from flames and sources of ignition, to prevent personal injury.
   - *Alcoholic beverage stains* Alcoholic beverage stains turn brown the longer they stay on the fabric. As soon as the stain appears on the fabric, start treating the stain immediately. Wash or soak the stain in cold water, then wash the garment.
   - *Blood stains* To remove blood stains, rinse or soak garment in cold water with an enzyme pre-soak product. You can use diluted chlorine bleach on white fabrics, if necessary. For colored fabrics, use a powdered oxygen-type bleach. Then, wash the garment.

- *Chewing gum stains* To remove chewing gum, use ice on the stain to make the chewing gum hard. Then, scrape most of it off the fabric. Next, use a non-flammable dry-cleaning solvent with a sponge to remove the excess chewing gum. Wash the garment.
- *Coffee or chocolate stains* To remove coffee or chocolate stains, soak the garment in warm or cold water. Next, make a paste of detergent mixed with hot water, and brush it on the stain. Wash the garment.
- *Milk product stains* To remove a milk product stain, use a non-flammable dry-cleaning solvent with a sponge. Wash the garment.
- *Antiperspirant and deodorant stains* To remove antiperspirant and deodorant stains, wash the garment with laundry detergent in the hottest water that is still safe for the fabric. If the stain remains on the fabric, then place the stain face down on a white towel, and treat the stain with a paste of ammonia and a powdered oxygen-type bleach. Let the paste stay on the stain for 30 minutes, then wash the garment in the hottest safe water for the fabric.
- *Fruit stains* To remove fruit stains, soak the stain in cool water. Do not use soap, it will set the stain. Wash the garment. If the stain remains, cover the stain with a paste made of a powdered oxygen-type bleach, a few drops of hot water, and a few drops of ammonia. Let the paste stay on the stain for about 15 to 30 minutes. Then, wash the garment.
- *Iron or rust stains* To remove iron or rust stains, apply some lemon juice mixed with salt. Then, place the garment in the sun. Alternatively, a commercial rust-removing solution can be used. Wash the garment.

YELLOWING IN FABRIC

Some of the causes of yellowing in fabrics are:
- Poor body soil removal
- Clothes washed in water treated with a water softener
- Hard water, minerals in the water, such as iron
- Body oils released into the garment
- The water supply might pick up the color of decaying vegetation

To remove body oils, the user will have to increase the amount of detergent, and use 150-degree Fahrenheit wash water. The user must also increase the frequency of using bleach in the wash.

To remove the yellowing from the garments that are washed in water from a water softener; the user will have to decrease the amount of detergent used, approximately to the point that the decreased amount will not affect the soil removal process. The user must also increase the frequency of using bleach in the wash.

Hard water and minerals in the water can be treated with a water conditioning apparatus. The user might have to drain the water heater, and flush the tank. Never use chlorine bleach to remove hard water stains or iron stains.

To remove body oils from the garment, use a paste made of detergent and water. Let it stay on the fabric for 15 to 30 minutes. Then, wash the garment.

To remove the yellowing caused by decaying vegetation, increase the amount of detergent, and bleach more often. White fabrics will respond very well to bleaching.

Fabric softener stains

Fabric softener stains are becoming more prevalent, because it is now being recommended that some fabric softeners can be used in the wash cycle, instead of the rinse cycle. These types of stains show up on synthetics, as well as cotton fabrics. They can be removed from the fabric by pretreating the stain with liquid detergent and following the washing procedures listed in the use and care manual.

Lint

Lint is cotton fiber that has broken away from the cotton garment. Lint likes to attach itself to synthetic fabrics. When this happens, the user often thinks that the washer is not performing properly. Therefore, to solve the problem of lint on synthetic fabrics, the user must sort the items before washing the clothes. For example:

1. The user must separate cottons from permanent press and knits.
2. The user must separate light colors from dark colors. Another cause of lint on the clothes is overwashing. This causes the clothes to wear out faster. To correct the overwashing problem, use only one minute of wash time per pound of dry laundry with normal soil. Any more time than this is a waste, and it usually does not get the laundry cleaner.

If the drain cycle is excessive, this too will cause lint to remain on the garments. Check for improper drain hose connections. For example:

1. Drain hose is too long; over 10 feet,
2. Drain hose is too high; over 5 feet, or
3. Is the drain hose kinked?

If excessive drain times still exist, then check the following:

1. Check the filter, located under the wash basket on most newer models.
2. Check to be sure that the pump is operating properly.
3. Check for any obstructions in the drain system.
4. Check for any obstructions within the water circulatory system of the washer.

AUTOMATIC WASHER MAINTENANCE

The interior is normally self-cleaning. However, there are times when you might have to remove objects from the inner basket. Clean the control panel and outer cabinet with a soft damp cloth. Do not use any abrasive powders or cleaning pads. Clean and inspect the interior of the underneath of the washer. Read the use and care manuals for the proper maintenance of your brands of washers.

REPAIR PROCEDURES

Each repair procedure is a complete inspection and repair process for a single washer component; containing the information you need to test a component that might be faulty; and to replace it, if necessary.

Washer timer

The typical complaints associated with washing machine timer failure are: 1. The cycle will not advance. 2. The washer won't run at all. 3. The washer will not fill. 4. The washer will not pump the water out. 5. The washer will not shut off.

1. *Verify the complaint* Verify the complaint by operating the washer through its cycles. Before you change the timer, check the other components controlled by the timer.

2. *Check for external factors* You must check for external factors not associated with the appliance. Is the appliance installed properly? Does the appliance have the correct voltage?

3. *Disconnect the electricity* Before working on the washer, disconnect the electricity. This can be done by pulling the plug out from the electrical outlet. Be sure that you only remove the washer plug. Or disconnect the electricity at the fuse panel or the circuit breaker panel. Turn off the electricity.

4. *Remove the console panel to gain access* Begin with removing the screws from the console panel to gain access to the timer (Fig. 10-2). Roll the console toward you.

5. *Test the timer* Remove the timer motor leads from the timer assembly. Test the timer motor by connecting the ohmmeter probes to the timer motor leads (Fig. 10-3). Set the range on the ohmmeter to R × 100. The meter should indicate between 2000 and 3000 ohms. Next, test the timer switch contacts, with the wiring diagrams configuration for the affected cycle. Place the meter probe on each terminal being tested, and turn the timer knob. If the switch contact is good, your meter will read continuity. If the timer motor measures suitably, then connect a 120-volt, fused service cord (Fig. 10-4) to the timer motor leads. Note: Connect the ground (common) wire test lead to the console ground wire.

    Be cautious whenever you are working with "live" wires. Avoid any shock hazards.

    If the motor does not operate, replace the timer. If the timer motor runs, but does not advance the cams, then the timer has internal defects and should be replaced.

6. *Remove the timer* To remove the timer, remove the timer mounting screws (Fig. 10-5). Remove the wire lead terminals from the timer. Mark the wires as to their location on the timer. Some timers have a disconnect block, instead of individual wires, which makes it easier to remove the timer wires. Turn the timer knob counterclockwise, to remove it from the timer shaft, and slide the indicator dial off the shaft.

7. *Install a new timer* To install a new timer, just reverse the disassembly procedure, and reassemble. Replace the wires on the timer. Reinstall the console panel, and restore the electricity to the washer. Test the washing machine cycles.

Water temperature selector switch

The typical complaints associated with the water temperature selector switch are: 1. Inability to select a different water temperature. 2. The consumer inadvertently selected the wrong water temperature.

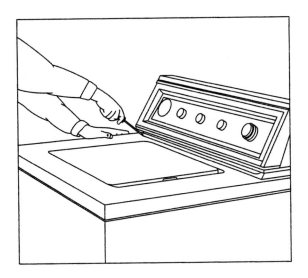

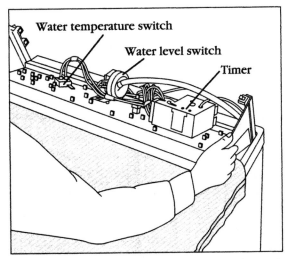

Water temperature switch

Water level switch

Timer

10-2 Remove the screws that hold the control console. Fold over the console to gain access.

1. *Verify the complaint* Verify the complaint by trying to select different water temperatures.
2. *Check for external factors* You must check for external factors not associated with the appliance. Is the appliance installed properly? Is there any physical damage to the component? Are the fill hoses connected to the hot and cold water supply correctly? Be sure that both the hot and cold water faucets are turned on.
3. *Disconnect the electricity* Before working on the washer, disconnect the electricity. This can be done by pulling the plug out from the electrical outlet. Be sure that you only remove the washer plug. Or disconnect the

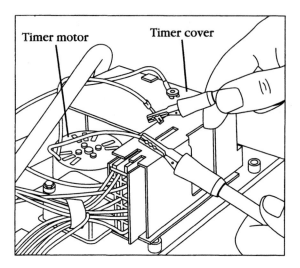

10-3
Checking the washer timer motor.

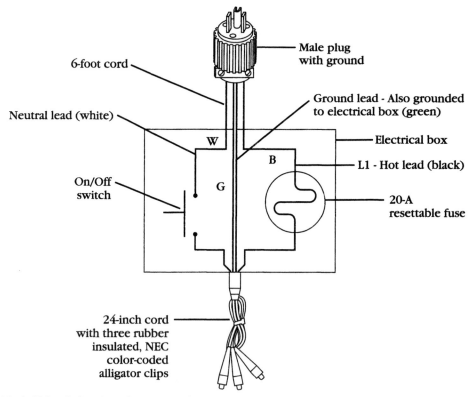

10-4 120-volt fused service test cord.

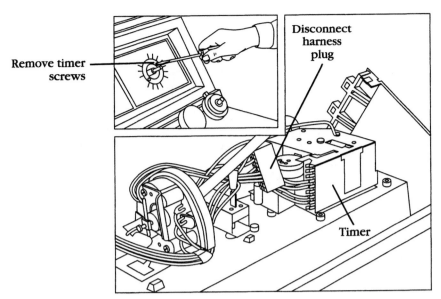

Remove timer screws

Disconnect harness plug

Timer

10-5 Removing the timer.

electricity at the fuse panel or at the circuit breaker panel. Turn off the electricity.

4. *Remove the console panel to gain access* Begin with removing the console panel to gain access to the water temperature selector switch (Fig. 10-2).

5. *Test the water temperature selector switch* To test the water temperature selector switch, remove all wires from the switch. Label the wires. Remember, you will have to identify the wires according to the wiring diagram, in order to reinstall them back onto the water temperature selector switch properly. Take your ohmmeter and check for continuity on the switch contacts. Press or turn the switch that coincides with the terminals that are being checked (Fig. 10-6). At this point, you have to use the wiring diagram to identify the switch contacts.

6. *Remove the water temperature selector switch* To remove the water temperature selector switch, remove the screws that hold the component to the console panel (Fig. 10-7).

7. *Reinstall the water temperature selector switch* To reinstall the water temperature selector switch, just reverse the disassembly procedure, and reassemble. Note: you will have to identify the wires according to the wiring diagram in order to reinstall them back onto the water temperature selector switch properly. Reinstall the console panel, and restore the electricity to the washer. Test the washing machine water temperature cycles.

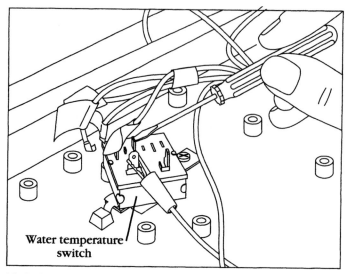

10-6 Checking the water temperature switch.

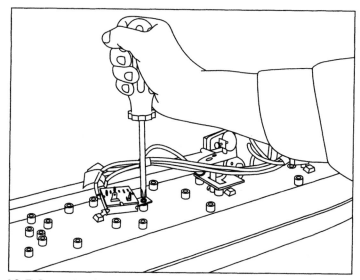

10-7 Removing the water temperature switch.

Water valve

The typical complaints associated with water valve failure (Fig. 10-8) are: 1. The washer will not fill with water. 2. The washer will overfill, and leak onto the floor. 3. When the washer is off, water still enters the tub.

    1. Verify the complaint Verify the complaint by operating the washer through its cycles. Listen carefully, and you will hear whether the water is entering the washer.

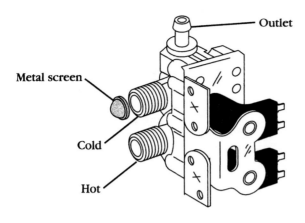

Metal screen

Outlet

Cold

Hot

10-8
A typical water valve used in automatic washers.

2. *Check for external factors* You must check for external factors not associated with the appliance. Is the appliance installed properly? Does the appliance have the correct voltage? Is the water turned on?

3. *Disconnect the electricity* Before working on the washer, disconnect the electricity. This can be done by pulling the plug from the electrical outlet. Be sure that you only remove the washer plug. Or disconnect the electricity at the fuse panel or at the circuit breaker panel. Turn off the electricity.

4. *Gain access to the water valve* Turn off the water supply to the water valve. To access the water valve, the rear panel must be removed. Disconnect the fill hoses from the inlet end of the water valve (Fig. 10-9). Next, remove the screws that hold the water valve to the chassis of the washer (Fig. 10-10).

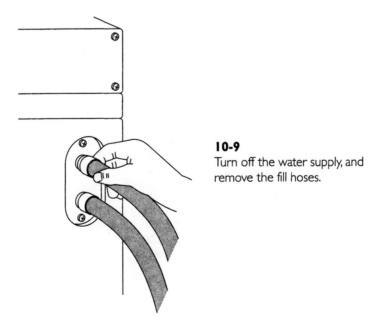

10-9
Turn off the water supply, and remove the fill hoses.

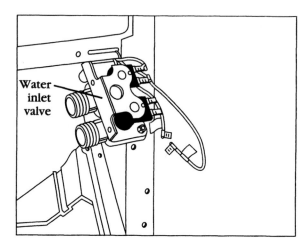

10-10
Remove the wires from the solenoid coil.

5. *Test the water valve* In order to check the solenoid coils on the water valve, remove the wire leads (and label them) that connect to the coils from the wire harness (Fig. 10-10). These are slide-on terminal connectors attached to the ends of the wire. Just pull them off. Set the ohmmeter on R × 100, and attach the probes to the terminals of one of the solenoid coils (Fig. 10-11). The meter should read between 500 and 2000 ohms. Repeat this test for the second solenoid coil.

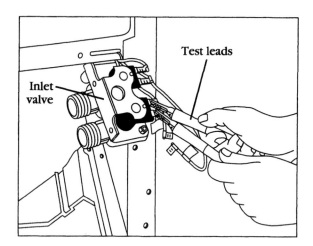

10-11
Attaching test leads to the solenoid coil on the water valve.

To test the fill rate of the water valve, just reverse the disassembly procedure, and reassemble the water valve. The rear panel does not have to be reinstalled for this test. Attach the 120-volt fused service cord; including the ground wire test lead to the cabinet ground (Fig. 10-4); to the water valve solenoid coil (Fig. 10-11). Then, energize the solenoid coil to allow water to enter the tub, and to check the flow rate of the water valve (Table 10-1). This step is repeated for each solenoid coil. If, when you energize the

Table 10-1

Water fill rate for a typical water valve

Water input P.S.I.	Gallon per minute one side of valve	Gallon per minute both sides of valve
20	3.7	4.5
30	4.6	5.5
40	5.3	6.2
50	5.6	6.3
60	5.8	6.7
80	6.6	7.0
100	7.0	7.2
120	7.2	7.4
140	7.0	7.3
160	6.8	7.0

water valve, no water enters the washer tub, replace the water valve. If the water valve checks correctly, then check the timer and the wiring harness.

6. *Remove the water valve* To remove the water valve, follow the instructions in step 4. Remove the water outlet hose from the water valve.
7. *Install a new water valve* To install the new water valve, just reverse the disassembly procedure, and reassemble. Reconnect the wire leads to the solenoid coils. After the installation of the new valve, turn on the water supply and check for water leaks. If none are found, reinstall the rear panel and restore the electricity to the washer. Set the timer and the water temperature control settings to operate the washer through its cycles.

Washer motor

The typical complaints associated with motor failure are: 1. Fuse is blown, or the circuit breaker trips. 2. Washer fills up with water, and the motor will not run.

1. *Verify the complaint* Verify the complaint by operating the washer through its cycles. Listen carefully, and you will hear if there are any unusual noises, or if the circuit breaker trips.
2. *Check for external factors* You must check for external factors not associated with the appliance. Is the appliance installed properly? Does the appliance have the correct voltage?
3. *Disconnect the electricity* Before working on the washer, disconnect the electricity. This can be done by pulling the plug from the electrical outlet. Be sure that you only remove the washer plug. Or disconnect the electricity at the fuse panel or at the circuit breaker panel. Turn off the electricity.
4. *Gain access to the motor* To access the motor, the back panel must be removed (Fig. 10-12). The back panel is held on with screws. Remove the screws and remove the panel.

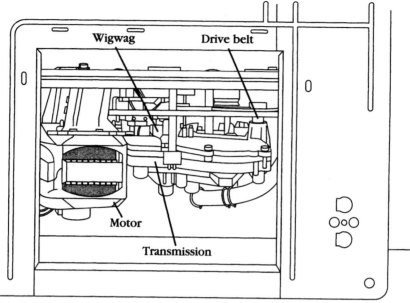

10-12 Removing the screws that hold the back panel.

5. *Disconnect the motor wire leads* Disconnect the motor wire leads from the wiring harness. Set the ohmmeter on R × 1, and attach the probes to the motor lead wires (Fig. 10-13). Refer to the wiring diagram for the common, start, and run motor winding leads identification. Test these for continuity, from the common wire lead to the run winding. Then, test for continuity from the common wire lead to the start winding. Next, test for continuity from the start winding to the run winding. To test for a grounded winding in the

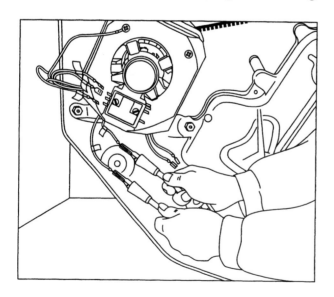

10-13
Checking motor windings for continuity.

motor, take the ohmmeter probes and test from each motor wire lead, to the motor housing (Fig. 10-14). The ohmmeter will indicate continuity if the windings are grounded. If the motor has no continuity between the motor windings, then replace the motor. If the motor checks OK, then check the timer, and motor relay (if the model that you are repairing has one).

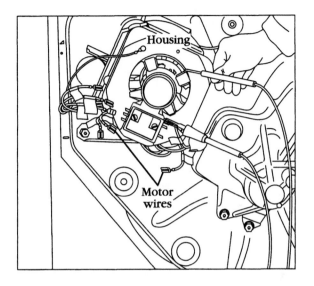

10-14
Checking the motor for ground.

6. *Remove the motor* To remove this type of motor, you must first loosen the two nuts that hold the motor support bracket (Fig. 10-15). Then, slide the assembly to disengage the belt from the pulley. Next, remove the four nuts that hold the motor to the motor support bracket (Fig. 10-16). Remove the motor from the washer. Remember to remove any remaining wires from the motor, and label them. Remove the pulley from the motor after loosening the set screw.

7. *Install the new motor* To install the new motor, just reverse the disassembly procedure, and reassemble. To adjust the belt, refer to the "drive belt" section of this chapter (step 5). Restore the electricity to the washer, and test the motor. If the motor is working, reinstall the back panel.

Capacitor

The typical complaints associated with capacitor failure are: 1. Fuse is blown, or the circuit breaker trips. 2. Washer motor will not run. 3. Motor has burning smell. 4. Motor will try to start, and then shuts off.

1. *Verify the complaint* Verify the complaint by operating the washer. Listen carefully, and you will hear if there are any unusual noises, or if the circuit breaker trips. If you smell something burning, immediately turn off the washer, and pull the plug.

2. *Check for external factors* You must check for external factors not associated with the appliance. Is the appliance installed properly? Does the appliance have the correct voltage?

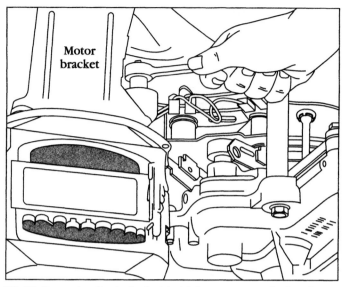

10-15 Removing the bolts that hold the motor on the bracket.

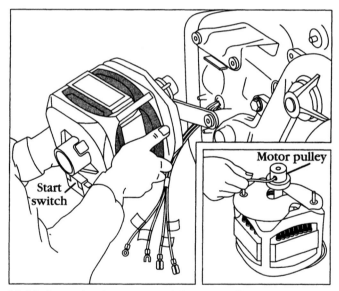

10-16 Removing the motor and then the pulley.

3. *Disconnect the electricity* Before working on the washer, disconnect the electricity. This can be done by pulling the plug from the electrical outlet. Be sure that you only remove the washer plug. Or disconnect the electricity at the fuse panel or at the circuit breaker panel. Turn off the electricity.

4. *Gain access to the capacitor* Some models have the capacitor mounted on the motor, and some are mounted to the cabinet interior, in the rear of the

machine. Access might be achieved through the front or rear panels, depending on which model you are working. Do not touch the capacitor until it's discharged.

Warning: A capacitor will hold a charge indefinitely; even when it is not currently in use. A charged capacitor is extremely dangerous. Discharge all capacitors immediately, any time that work is being conducted in their vicinity. Redischarge after repowering the equipment, if further work must be done.

5. *Test capacitors* Before testing the capacitor, it must be discharged. Use a screwdriver, with an insulated handle, to discharge the capacitor by shorting it across both terminals. Remove the wire leads, one at a time, with needle nose pliers. Set the ohmmeter on the highest scale, place one probe on one terminal and the other probe on the other terminal (Fig. 10-17). Observe the meter action. While the capacitor is charging, the ohmmeter will read nearly zero ohms for a short period of time. Then, the ohmmeter reading, will slowly begin to return toward infinity. If the ohmmeter reading deflects to zero, and does not return to infinity, the capacitor is shorted; and it should be replaced. If the ohmmeter reading remains at infinity, and does not dip toward zero, the capacitor is open; and it should be replaced.

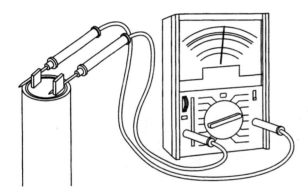

10-17
Placing ohmmeter test leads on the capacitor terminals.

6. *Remove the capacitor* Remove the capacitor from its mounting bracket.
7. *Install a new capacitor* To install the new capacitor, just reverse the disassembly procedure, and reassemble. Note: a capacitor is rated by its working voltage (WV, or WVac), and by its storage capacity in microfarads (μF). Always replace a capacitor with one having the same voltage rating, and the same (or up to 10% greater) microfarad rating.

Drive belt

The typical complaints associated with belt failure are: 1. Washer will not agitate. 2. Washer will not spin. 3. Washer motor spins freely. 4. Smells like something is burning.

1. *Verify the complaint* Verify the complaint by operating the washer in the spin cycle. Listen carefully, and you will hear and see if the inner basket is turning, or if the circuit breaker trips.

2. *Check for external factors* You must check for external factors not associated with the appliance. Is the appliance installed properly? Does the appliance have the correct voltage?

3. *Disconnect the electricity* Before working on the washer, disconnect the electricity. This can be done by pulling the plug out from the electrical outlet. Be sure that you only remove the washer plug. Or disconnect the electricity at the fuse panel or at the circuit breaker panel. Turn off the electricity.

4. *Gain access to the belt* You must gain access to the belt, either by removing the back or the front panel, or by tilting the washer, or by laying it onto its back; depending on which model washer you are working. The back panel (or the front panel) is usually held on with two screws. Remove the screws, and remove the panel.

5. *Adjust the belt* Before adjusting the belt, use your finger and press on the belt; it should only deflect about ¼ inch. To adjust the belt (Fig. 10-15), loosen the motor bracket nut just enough to move the bracket. Take hold of the motor bracket, and pull against the belt just enough to take up the slack in it, and to properly re-tension it. If you are unable to adjust the tension, or if the belt is worn, replace the belt. Some models have more than one belt. One is for the drive system that is attached to the motor and the transmission pulleys. The other belt is for the water pump. This belt is attached to the motor pulley and to the water pump pulley (Fig. 10-18). To adjust the water pump belt in this type of washer, just loosen the pump mounting screws and adjust to obtain the correct tension; about ¼-inch deflection. There are even some models that use a direct drive system (which has no belts) to drive the motor, transmission and pump (Fig. 10-19). This type of washer has the motor and the water pump attached to the transmission with retaining clips. Between the motor shaft and the transmission, there is a coupling. To replace this coupling, remove the retaining clips that secure the water pump and the motor. Install the new coupling, and reattach the motor and the water pump.

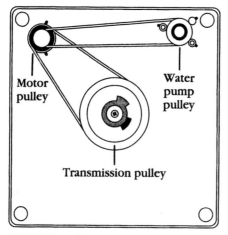

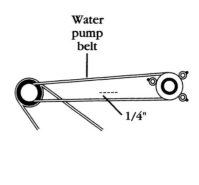

10-18 View from underneath of the washer base. Check belt deflection.

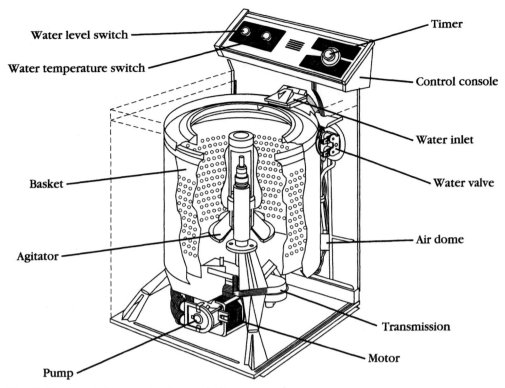

Water level switch

Water temperature switch

Timer

Control console

Water inlet

Basket

Water valve

Agitator

Air dome

Transmission

Motor

Pump

10-19 The direct drive washer has no belts.

6. *Replace the drive belt* To replace the drive belt on this type of washer, you must remove the flexible pump coupling (Fig. 10-20). Next, loosen the motor bracket nuts, and slide the motor forward to take the tension off the belt. Then, remove the belt from the motor pulley and the transmission pulley (Fig. 10-21). Install the new belt on the transmission pulley and the clutch pulley. Be sure that the belt is in the pulley grooves. Next, adjust the belt tension, and tighten the motor bracket nuts. Reinstall the flexible pump coupling and clamps, be sure that the coupling is not twisted, and that it is seated on the pump and clutch pulleys. The drive belt tension on this type of washer should be approximately ½ inch when deflected.

7. *Test the washer* You are now ready to test the washer. Begin the wash cycle with a full load of laundry in the basket. Check the agitate and the spin cycles. If these check, reinstall outer panels. If not, readjust the belt tension.

Water level control

The typical complaints associated with the water level control are: 1. Water flowing over the top of the tub. 2. Tub does not fill to the proper level selected. 3. Washer will not agitate. 4. Washer will not spin.

1. *Verify the complaint* Verify the complaint by trying to select different water levels when operating the washer through its cycles.

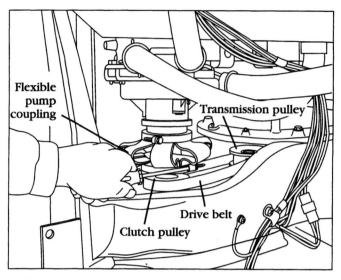

Flexible pump coupling

Transmission pulley

Drive belt

Clutch pulley

10-20 Loosening the screw on each clamp and removing the flexible coupling.

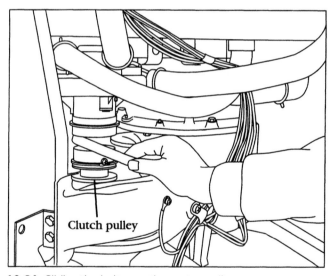

Clutch pulley

10-21 Sliding the belt over the motor pulley.

2. *Check for external factors* You must check for external factors not associated with the appliance. Is the appliance installed properly? Does the appliance have the correct voltage? Is there any physical damage to the component? Is the plastic line connected to the water level control and air dome? Check to be sure that the water is turned on all the way.

3. *Disconnect the electricity.* Before working on the washer, disconnect the electricity. This can be done by pulling the plug from the electrical outlet.

Be sure that you only remove the washer plug. Or disconnect the electricity at the fuse panel, or at the circuit breaker panel. Turn off the electricity.

4. *Remove the console panel to gain access* Begin with removing the console panel, to gain access to the water level control (Fig. 10-2).

5. *Test the water level control* To test the water level control switch, remove and label all of the wires from the switch (Fig. 10-22). Remember, you will have to identify the wires according to the wiring diagram, in order to reinstall them back on the water level control switch properly. With the washer empty, use your ohmmeter (set on R × 1) and test it for continuity on the switch contacts numbered 1 and 2 (Fig. 10-23). If you have continuity, this means that the water valve will be energized, allowing the water to enter the tub. Now, test for continuity between contacts 1 and 3; the ohmmeter should read no continuity. Reconnect the wires to the water level switch. Plug in the washer, and start the wash cycle. Let the console rest on top of the washing machine for this test. Be careful not to touch any live wires, or to short them to the washer chassis while the test is being performed.

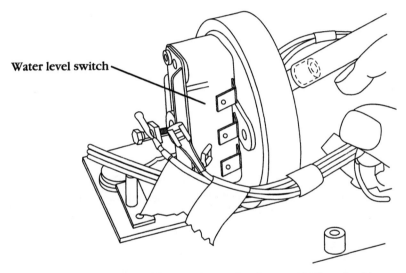

Water level switch

10-22 Removing the wires from the water level control before checking continuity.

As the water level rises in the tub, it forces air through the air dome and up the plastic tube to the water level control. The pressure that is exerted on the water level control's diaphragm will trip the water level switch, from empty to full, which will start the agitation cycle. When the agitation cycle begins, turn off the washer. Pull the power plug from the wall socket to ensure that power has been removed.

At this point, test for continuity again between the contacts numbered 1 and 2. The ohmmeter should read no continuity. Next, test for continuity between contacts numbered 1 and 3. The ohmmeter will read continuity. If the water level switch checks out, then check the plastic hose that goes

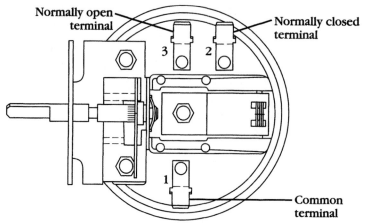

Normally open terminal

Normally closed terminal

3 2

1

Common terminal

10-23 Adjustable water-level control terminal identification for this model only.

from the air dome to the water level control. Ensure that there are no holes or cracks in the line. If the switch checks bad, replace it. Remember this, never blow into the water level control switch to activate it. Why? You might activate the switch, but it will not prove that the switch will activate (at lesser pressures) at the proper water level setting selected.

6. *Remove the water level control* To remove the water level control, remove the wires, then remove the screws that hold the component to the console panel (Fig. 10-24). Next, remove the plastic hose.

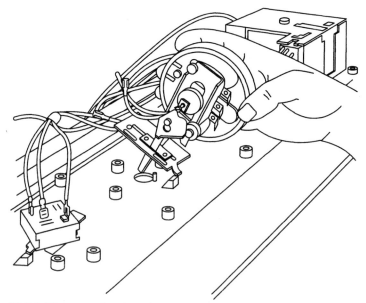

10-24 Disconnecting the wires and removing the screws that hold the control to the console.

7. *Reinstall the water level control.* To reinstall the water level control, just reverse the disassembly procedure, and reassemble. Remember, you will have to identify the wires according to the wiring diagram in order to properly reinstall them back on the water level control.

Lid switch

The typical complaints associated with the lid switch are: 1. Washer will not spin. 2. Washer will not agitate.

1. *Verify the complaint* Verify the complaint by closing the washer lid, and turning the timer dial to start the spin cycle.

2. *Check for external factors* You must check for external factors not associated with the appliance. Is the appliance installed properly? Does the appliance have the correct voltage? Is there any physical damage to the component?

3. *Disconnect the electricity* Before working on the washer, disconnect the electricity. This can be done by pulling the plug from the electrical outlet. Be sure that you only remove the washer plug. Or disconnect the electricity at the fuse panel, or at the circuit breaker panel. Turn off the electricity.

4. *Gain access to the lid switch* If the washer's top snaps in place, tape the lid shut. Use a putty knife to release the spring clips in each corner, and lift the top. If the spring clips won't release, open the lid, pull the top toward you, and lift (Fig. 10-25). Raise the washer top to gain access to the lid switch (Fig. 10-26).

On some models, the top is held down with two screws that are secured from underneath the top. The front of the cabinet is secured in place with two screws; take them out and remove the front panel. Then, remove the screws that hold the top in place.

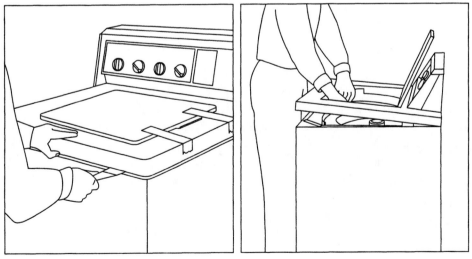

10-25 Taping the lid closed. Then pry the top open, and lift the top.

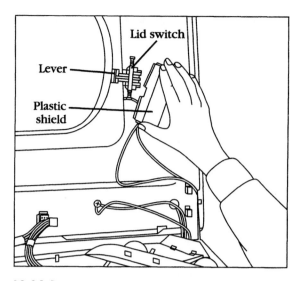

10-26 Removing the plastic shield to gain access to the lid switch.

5. *Test the lid switch* To test the lid switch, remove the two wires from the switch (Fig. 10-27). With a continuity tester, test for continuity between the two terminals of the switch while moving the lid switch lever. If the switch fails these tests, replace the switch.

6. *Replace the lid switch* Remove the screws that hold the switch in place (Fig. 10-28). Install the new switch, and connect the wires on the switch terminals. Then, reconnect the electricity and test the washer.

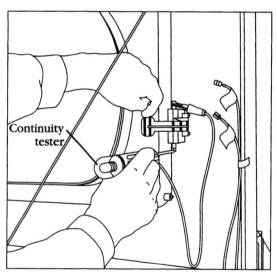

10-27 Removing the wires to check for continuity of the switch.

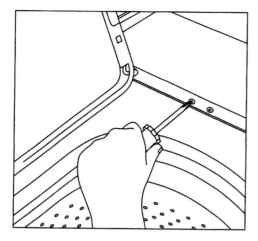

10-28
Removing the two screws that
hold the lid switch in place.

Water pump

The typical complaints associated with the water pump are: 1. Washer will not drain the water out. 2. Smells like something is burning. 3. The water in the washer will not recirculate.

1. *Verify the complaint* Verify the complaint by operating the washer.

2. *Check for external factors* You must check for external factors not associated with the appliance. Is the appliance installed properly? Does the appliance have the correct voltage?

3. *Disconnect the electricity* Before working on the washer, disconnect the electricity to the washer. This can be done by pulling the plug from the electrical outlet. Be sure that you only remove the washer plug. Or disconnect the electricity at the fuse panel or at the circuit breaker panel. Turn off the electricity.

4. *Access and remove the water pump* In order to gain access to the water pump in this type of washer, the cabinet must be removed. Remove the screws that hold the control panel in place. Next, lift the control panel up. Then, remove the clips from each side that hold the cabinet in place. Pull back on the cabinet, and move it out of the way (Fig. 10-29). With no water in the tub, remove the hoses from the pump ports (some water will spill out of the pump). Wipe the water up immediately. Check the water pump ports for any obstructions. If no obstructions are found, then disconnect the two clamps that hold the pump in place. Remove the pump.

   To gain access to the water pump in this type of washer, remove the back panel. With no water in the tub, remove the hoses from the pump ports (some water will spill out of the pump). Wipe the water up immediately. If obstructions are found, remove the obstructions, reconnect the hoses, and test for proper operation. If the unit is still not operating properly or if the obstruction could not be removed, then the pump must be removed. Next, remove the flexible pump coupling (Fig. 10-30). Now, remove the bolts that hold the pump in place (Fig. 10-31). Remove the pump.

5. *Install a new water pump* To install the new pump, just reverse the disassembly procedure, and reassemble. Test the washer for any water leaks.

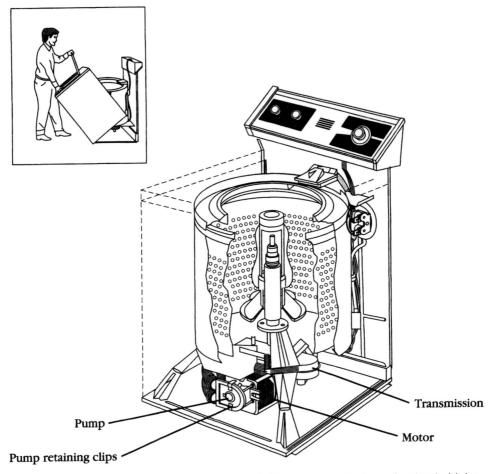

Transmission

Pump

Motor

Pump retaining clips

10-29 Removing the screws that hold the console. Then remove the two clips that hold the cabinet in place and remove the cabinet.

Inner basket

The typical complaints associated with the inner basket are: 1. Basket will not spin. 2. Water will not drain out of the tub. 3. Washer damages the clothing.

1. *Verify the complaint* Verify the complaint by inspecting and operating the washer.

2. *Check for external factors* You must check for external factors not associated with the appliance. Is the appliance installed properly? Does the appliance have the correct voltage?

3. *Disconnect the electricity* Before working on the washer, disconnect the electricity. This can be done by pulling the plug from the electrical outlet. Be sure that you only remove the washer plug. Or disconnect the electricity at the fuse panel, or at the circuit breaker panel. Turn off the electricity.

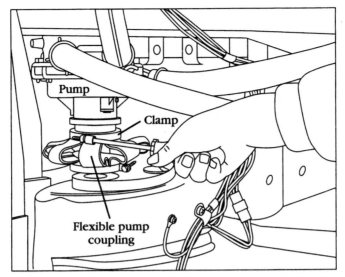

10-30 Loosen the screw on each clamp, and remove the flexible coupling.

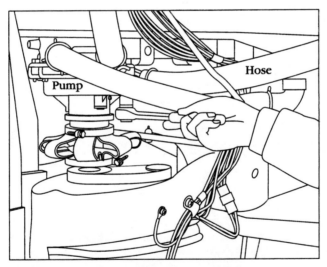

10-31 Removing the pump hold down bolts.

4. *Gain access to inner basket* To gain access to the inner basket, you must first lift the top of the washer (Fig. 10-25) and disconnect the fill hose from the water inlet, located on top of splash guard (Fig. 10-32). Remove the splash guard by removing the clips that hold it to the outer tub. Now that you have more room to work with, remove the agitator. On some models, the agitator is held down with a cap and stud assembly, or the agitator might just snap onto the transmission shaft (Fig. 10-33). Next, use a

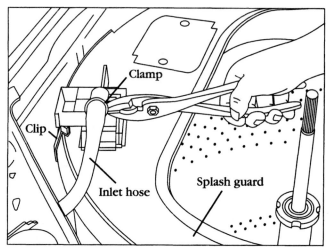

10-32 Removing the clamp from the inlet hose.

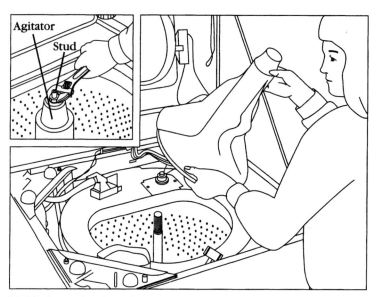

10-33 Removing the agitator.

spanner wrench to remove the locknut that holds the basket in place (Fig. 10-34). On some models, the inner basket is held in place with four bolts. If so, remove the bolts. Next, lift the inner basket out of the tub. Inspect the outer tub for debris and rust. Be sure that the tub drain opening is clear of debris.

5. *Reinstall inner basket* To reinstall in inner basket, just reverse the disassembly procedure, and reassemble. One important note, do not

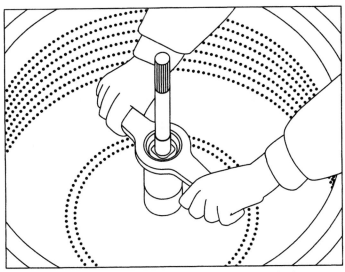

10-34 Always use a spanner wrench to remove the locknut.

overtighten the locknut. Just tighten the locknut enough to secure the basket in place. If you overtighten the locknut, the washer will not work properly.

6. *Test washer* After you have reassembled the inner basket, splash guard, and agitator, test the washer. Check for leaks around the top rim of the outer tub. Check for agitation and spin, with a full load of clothes.

Hoses

The typical complaints associated with the hoses are: 1. Washer leaks water. 2. Water will not pump out. 3. Kinked or plugged hoses.

1. *Verify the complaint* Verify the complaint by inspecting and operating the washer.

2. *Check for external factors* You must check for external factors not associated with the appliance. Is the appliance installed properly? Does the appliance have the correct voltage?

3. *Disconnect the electricity* Before working on the washer, disconnect the electricity. This can be done by pulling the plug from the electrical outlet. Be sure that you only remove the washer plug. Or disconnect the electricity at the fuse panel, or at the circuit breaker panel. Turn off the electricity.

4. *Remove defective hose* To remove the defective hose, the tub must be empty of water. Loosen the clamps that hold the hose in place. On some models, the manufacturer uses a snap ring clamp. To remove this type of clamp, just squeeze the ends together (Fig. 10-35).

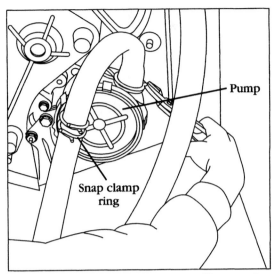

10-35 Removing the snap ring clamp. If you are just replacing a part, always check the hose for any cracks where the clamp was resting on the hose.

Clutch

The typical complaints associated with the clutch are: 1. Washer will not agitate. 2. Washer will not spin.

1. *Verify the complaint* Verify the complaint by operating the washer.

2. *Check for external factors* You must check for external factors not associated with the appliance. Is the appliance installed properly? Does the appliance have the correct voltage?

3. *Disconnect the electricity* Before working on the washer, disconnect the electricity. This can be done by pulling the plug from the electrical outlet. Be sure that you only remove the washer plug. Or disconnect the electricity at the fuse panel, or at the circuit breaker panel. Turn off the electricity.

4. *Gain access to the motor* To gain access to the motor, the back panel must be removed. The back panel is held on with screws. Remove the screws, and remove the panel.

5. *Remove the motor and clutch* In this type of washer, to remove the motor and clutch assembly, you must first disconnect the wires of the motor from the wiring harness. Then, remove the flexible pump coupling (Fig. 10-30). Remove the belt from the clutch. Once the motor is isolated, remove the three bolts that hold the motor mounting plate to the suspension. Now, remove the motor assembly from the washer. You are now ready to remove the clutch assembly. Remove the clutch drive plate by removing the roll pin. This is accomplished by using a drive pin tool on the roll pin (Fig. 10-36) and hitting it with a hammer. On a stubborn clutch drive plate, the use of a wheel puller tool and a horseshoe collar tool, will make it easier to remove (Fig. 10-37). Disassemble and remove the clutch assembly, and replace with a new assembly.

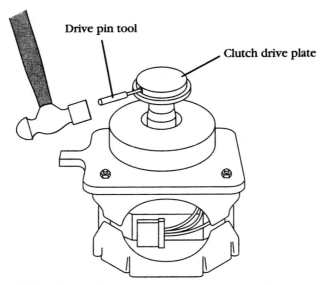

Drive pin tool

Clutch drive plate

10-36 When hitting the drive pin tool, be careful not to damage the clutch drive plate.

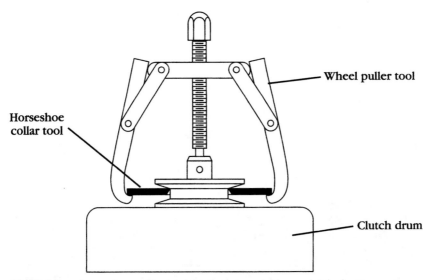

Wheel puller tool

Horseshoe collar tool

Clutch drum

10-37 Use a horseshoe collar tool to remove a stubborn clutch drum.

6. *Install new clutch assembly* To install the clutch assembly, just reverse the disassembly procedure, and reassemble. Adjust the belt tension.
7. *Test the washer* Test the washer operation by running the washer through a cycle. Be sure that you have some clothes in the washer when testing it.

Transmission

The typical complaints associated with the transmission are: 1. Washer will not agitate. 2. Washer will not spin.

1. *Verify the complaint* Verify the complaint by operating the washer.
2. *Check for external factors* You must check for external factors not associated with the appliance. Is the appliance installed properly? Does the appliance have the correct voltage?
3. *Disconnect the electricity* Before working on the washer, disconnect the electricity. This can be done by pulling the plug from the electrical outlet. Be sure that you only remove the washer plug. Or disconnect the electricity at the fuse panel or at the circuit breaker panel. Turn off the electricity.
4. *Remove inner basket.* To remove the inner basket, see the section labeled: Inner basket.
5. *Remove the transmission boot* To remove the transmission boot, loosen the ring clamps and lift the boot off the transmission and the outer tub (Fig. 10-38). Examine the boot, if it is damaged, replace it. If the ring clamps are rusted, replace them.
6. *Remove the transmission* Remove the six bolts that hold the transmission to the washer's suspension. Remember the position that the transmission is in before you remove it. This will help you later for the reinstallation of the new transmission. Next, reach in and remove the drive belt from the transmission pulley. Lift the transmission out of the washer (Fig. 10-39).
7. *Install the transmission* To reinstall the transmission, just reverse the disassembly procedure, and reassemble. Do not forget to reinstall the belt.
8. *Test the washer* Test the washer operation by running the washer through a cycle. Be sure that you have some clothes in the washer when testing it.

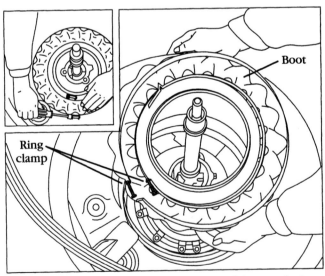

10-38 Loosen the clamps, and lift the boot off. Inspect the boot for holes.

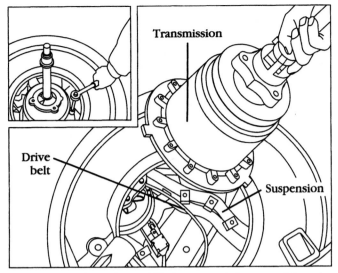

10-39 Removing bolts, and lift the transmission out of the tub.

DIAGNOSTIC CHARTS

The following diagnostic flow charts will help you pinpoint the likely causes of the problem (Figs. 10-40, 10-41, and 10-42).

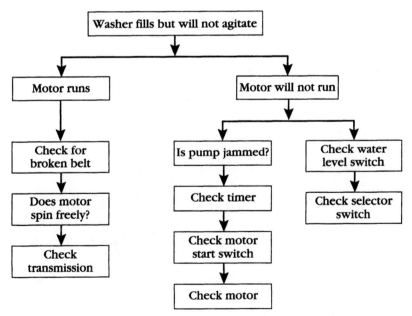

10-40 Diagnostic flow chart: washer fills, but will not agitate.

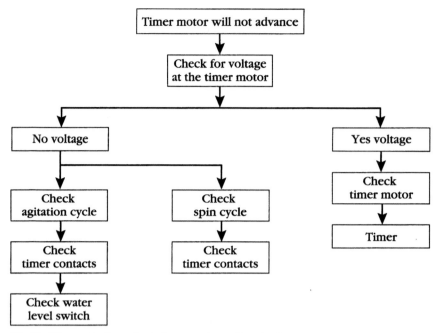

10-41 Diagnostic flow chart: timer motor will not advance.

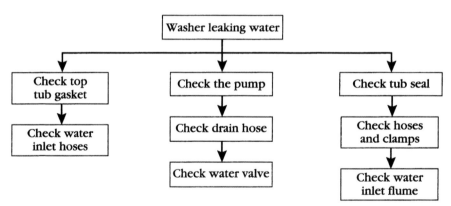

10-42 Diagnostic flow chart: washer leaking water.

The wiring diagrams in this chapter are used as examples only. You must refer to the wiring diagram on the washer that you are servicing. Figure 10-43 depicts an actual wiring schematic. Figure 10-44 depicts an actual ladder diagram of Fig. 10-43. A ladder diagram is generally easier to read.

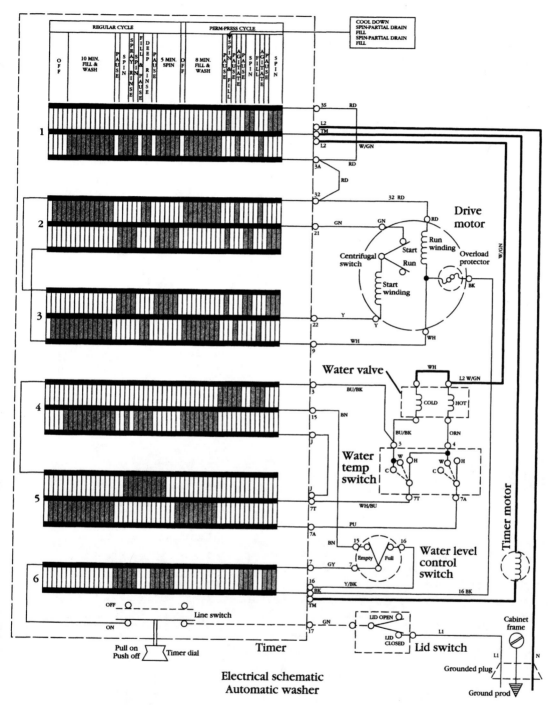

10-43 Automatic washer electrical schematic.

Ladder diagram
Automatic washer

10-44 Automatic washer ladder diagram.

11

Automatic dryers

Automatic electric dryers are not complicated to repair. The more that you know about the electrical, mechanical, and operational basics of the dryer, the easier it is to solve the problem. Some models are designed to operate on 240 volts, but other models operate on 120 volts. To dry clothing, an automatic dryer must have drum rotation, electric heat, and air circulation. This chapter will provide the basic skills needed to diagnose and to repair automatic dryers. Figure 11-1 identifies where the components are located within the automatic dryer.

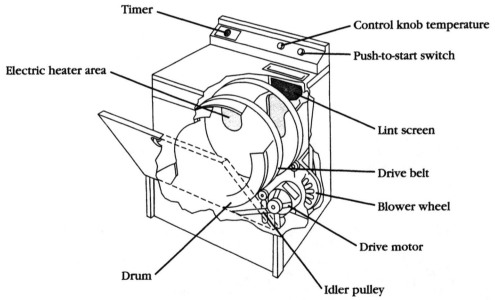

11-1 Illustrates component location in a typical automatic dryer.

PRINCIPLES OF OPERATION

The clothes are placed into the dryer according to the manufacturer's recommendation for the proper loading of the dryer. Next, the proper cycle is selected, and the dryer start button is pressed. The combination of the timer, the switches, and the thermostats regulates the air temperature within the drum and the duration of the drying cycle. During the drying cycle, room air is pulled into the dryer drum from the lower rear (or sometimes the front) of the cabinet, depending on which model the consumer owns. The air is pulled through the heater, drum, lint screen and down through the lint chute, through the fan housing, and it is then pushed out of the exhaust duct. The drive motor, blower wheel, belt, and pulleys cause the drum to turn, and the air to move through the dryer. The belt wraps around the drum, motor pulley, and idler pulley. The blower wheel is secured to one end of the drive motor shaft. As the drive motor turns, the drum rotates, moving the clothes. At the same time, the blower wheel turns, moving the air; and the heating element will cycle on and off, according to the temperature selected.

Functions and cycles

Electric automatic dryers use three basic functions to operate:
1. Heat is supplied by a resistance-type heating element.
2. Air is drawn into the dryer. It is heated and circulated through the tumbling clothes. Then, the warm, moisture laden air is drawn through the lint screen, and it is vented through the duct system to the outside.
3. Tumbling of the clothes is accomplished with a motor that drives a belt, that rotates the drum.
   - *Timed dry cycle* The timed dry cycle is controlled by the amount of time selected on the timer. The temperature in the dryer is controlled by a thermostat, which turns the heating element on and off throughout the timed cycle.
   - *Automatic dry cycle* The automatic dry cycle is not controlled by the timer. This cycle is controlled by the cycling thermostats. Heat is supplied to dry the clothes, and it will continue until the temperature in the drum reaches the selected cutout setting of the thermostat. When the thermostat is satisfied, the heater shuts off, and the timer motor is activated. But, there are variables available that can control the cycle, which will cause the cycling of the thermostat before the end of the cycle.
   - *Permanent press and knit cycle* The permanent press and knit cycle are controlled by the amount of time selected on the timer. The temperatures of these cycles are controlled by the temperature rating of the cycling thermostats that are located within the cabinet, on the exhaust duct, and in the air supply. On some models, the user can select the desired type of heat setting with the temperature selector switch, located on the control panel.
   - *Air dry* The air dry cycle is controlled by the amount of time selected on the timer. This cycle, uses the air to dry the clothes. The heating element is not used at all during this cycle.

SAFETY FIRST

Any person who cannot use basic tools, or follow written instructions, should *not* attempt to install, maintain, or repair any automatic dryers. Any improper installation, preventive maintenance, or repairs could create a risk of personal injury or property damage.

If you do not fully understand the installation, preventive maintenance, or repair procedures in this chapter; or if you doubt your ability to complete the task on the automatic dryer; then please call your service manager.

The following precautions should also be followed:

1. Never bypass (or interfere with) the operation of any switch, component, or feature.
2. The dryer exhaust should be vented properly. Never exhaust the dryer into a chimney, a common duct, an attic, or a crawl space.
3. Be careful of sharp edges when working on the dryer.
4. The dryer produces combustible lint, and the area should be kept clean.
5. Never remove any ground wires from dryer; nor the third (grounding) prong from the service cord.
6. Never use an extension cord to operate a dryer.
7. The wiring used in dryers is made with a special heat-resistant insulation. Never substitute it with ordinary wire.

Before continuing, take a moment to refresh your memory of the safety procedures in Chapter 2.

AUTOMATIC DRYERS IN GENERAL

Much of the troubleshooting information in this chapter covers automatic electric dryers in general, rather than specific models, in order to present a broad overview of service techniques. The pictures and illustrations that are used in this chapter are for demonstration purposes to clarify the description of how to service dryers. They in no way reflect upon a particular brand's reliability.

Electrical requirements

Electrical dryers can be connected to an electrical power source in three ways:

1. Electrical cord and plug, which plugs into a 240-volt wall receptacle.
2. Directly wired to a fused electrical disconnect box with a built-in shut-off switch.
3. Directly wired to a circuit breaker or to a fuse panel. All dryers must be properly wired, grounded, and polarized, according to the manufacturer's installation instructions, and per all local codes and ordinances.

Location and installation of dryer

Listed are some general principles that should be followed when installating a dryer.

1. Locate the dryer where there is easy access to existing electrical lines.
2. Be sure that you observe all local codes and ordinances for the electrical, plumbing and venting connections.
3. The dryer should be installed and leveled on a firm floor, to minimize vibration during operation.
4. To reduce the risk of fire, never install a dryer on any type of carpeting.
5. For proper operation of the dryer, be sure that there is adequate make-up (new, replacement) air in the room where it is installed.
6. Do not install a dryer in an area where the make up air is below 45 degrees Fahrenheit. This will greatly reduce the drying efficiency, and increase the cost of operating the dryer.

7. Always follow the installation instructions which are provided with every automatic dryer model purchased. If the installation instructions are not available, order a copy from the manufacturer.
8. When installing a dryer in a mobile home, the dryer must have an outside exhaust. If the dryer exhaust goes through the floor, and the area under the mobile home is enclosed, the exhaust must terminate outside of the enclosed area.
9. If the dryer is to be installed in a recessed area, or closet area, follow the manufacturer's recommendation for its installation.
10. When relocating the dryer to a new location, test the voltage at the new location, and be sure that it matches the dryer voltage specifications, as listed on the nameplate (serial plate). Also, inspect the entire vent system for any obstructions.

Proper exhausting of the dryer

Proper exhausting instructions for the model being installed are available through the manufacturer. Each manufacturer has their own specifications for the size and the length of the ductwork needed to run their dryer properly. The maximum length of the exhaust system depends upon the type of duct, the number of elbows, and the type of exhaust hood used. Figure 11-2 illustrates a typical dryer exhaust installation. The following guide is recommended for the exhausting of a dryer:
1. Keep the duct length as short as possible.
2. Keep the number of elbows to a minimum, to minimize the air resistance.
3. Never reduce the diameter of round ductwork below four inches.
4. Install all exhaust hoods at least 12 inches above ground level.
5. The exhaust duct, and exhaust hood, should be inspected periodically, and cleaned, if necessary.

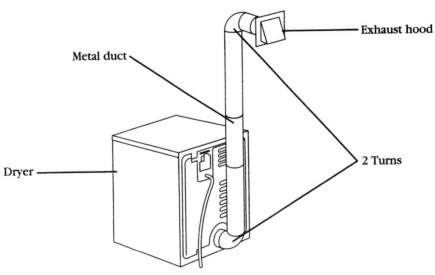

11-2 Typical dryer exhaust installation.

6. All duct joints should be taped. Never use screws to join the duct joints together. Screws protruding into the duct will cause lint build-up, and this will eventually clog the duct.

7. Never exhaust the dryer into any wall, ceiling, attic, or under a building. Accumulated lint could become a fire hazard, and moisture could cause damage.

8. If the exhaust duct, is adjacent to an air conditioning duct, the exhaust duct must be insulated to prevent moisture build-up in the duct.

STEP-BY-STEP TROUBLESHOOTING BY SYMPTOM DIAGNOSIS

In the course of servicing an appliance, don't overlook the simple things that might be causing the problem. Step-by-step troubleshooting, by symptom diagnosis, is based on diagnosing malfunctions, with possible causes arranged into categories relating to the operation of the dryer. This section is intended only to serve as a checklist; to aid you in diagnosing a problem. Look at the symptom that best describes the problem that you are experiencing with the dryer, then correct the problem.

Dryer will not run

1. Do you have the correct voltage at the dryer?
2. Test the door switch for continuity of the switch contacts.
3. Test for the continuity of the motor windings. Also check for a grounded motor.
4. Test the timer for the continuity of the switch contacts.
5. Test the push to start switch for the continuity of the switch contacts.
6. Check for broken or loose wiring. Also check the wire terminal connections that connect to the different components.

No heat

1. Do you have the correct voltage? The motor in a dryer runs on 120 volts. The heating element works on 240 volts, and on compact models 120/240 volts. Check voltage at the wall receptacle. Also, check the nameplate voltage rating for the model that you are working on.
2. Check for any in-line fuses that might have blown. Refer to the wiring diagram, located on the rear of the dryer, or behind the console panel.
3. Are the thermostats functioning properly?
4. Test the timer for the continuity of the switch contacts.
5. Test for continuity of the motor windings. Also, test for a grounded motor.
6. Test the heating element for continuity.
7. Test the temperature selector switch for continuity.
8. Check for broken or loose wiring.

Drum will not rotate

1. Check the drive belt. Is it broken or worn?
2. Check the idler pulley. Is the pulley seized up?

3. Check the drum support bearings. Be sure that they rotate freely.
4. Is the dryer overloaded with clothing?
5. Check for foreign objects that might be lodged between the drum and the bulkheads.

Dryer noisy

1. Check for loose components. Is everything secure, and in its proper place?
2. Check the idler pulley. Lint build-up can cause the idler pulley shaft to squeak.
3. Check the drive belt. Is the belt partially torn?
4. Check the drive motor. Is the pulley secured to the shaft? Also, check the fan assembly. Is it loose?
5. Check for lint, or foreign objects, lodged between the drum and the bulkheads.
6. Check the front and rear bearings.
7. Check the blower assembly. Be sure that the blower wheel is tight on the motor shaft end.
8. Is the dryer level?

Dryer runs and heats, but the clothes won't dry

1. Check for a defective thermostat. Use a thermometer to check the duct temperatures during the cycling of the dryer.
2. Check for a loose pulley or blower wheel. Inspect and tighten the blower wheel.
3. Check for restricted air flow.
4. Check exhaust vent.

Clothes too hot or fabric damage

1. Check the exhaust system for restrictions.
2. Checks for air leaks in the ductwork, and the drum front and rear seals.
3. Check the thermostats.
4. Check the temperature setting by customer.
5. Check the door latch and strike.
6. Inspect the interior of drum for any foreign objects protruding into the drum.

COMMON DRYING PROBLEMS

The drying of clothes will depend on several factors:
1. Clogged lint screen.
2. Exhaust duct to long, collapsed, or crimped.
3. If dryer air is exhausted into the area where the dryer is located. For example: the dryer is located inside a closet, and operating with the closet door closed.
4. Overloading of clothing in the dryer.
5. If dryer is located in an area where the temperature is below 45 degrees Fahrenheit.
6. Laundry washed in cold water will take slightly longer to dry.

7. The moisture content of certain clothes. Towels and denim retain more moisture, etc.
8. Wrong spin speed selected, for the type of laundry being washed. For example: Towels, denim, blankets, small rugs, etc.
9. Electric supply is less than what is needed to operate the dryer efficiently.
10. Load type (towels, bedspreads, jeans, etc.)

All of these must be taken into account, when diagnosing a complaint of clothes not drying properly.

Lint

Lint consist of fibers that have broken away from the fabric. It can collect inside the dryer cabinet and base, and create a fire hazard. This lint should be removed every one to two years by cleaning it out. Some lint can also collect in the door opening, the drum, the heater assembly, the blower assembly, and the duct assembly. This accumulation of lint should be removed periodically, when performing maintenance on the dryer; otherwise, it could create problems with the future use of the dryer. The lint screen should be cleaned every time the dryer is used. The duct system, that exhausts the air to the outside of the dwelling, should also be cleaned out every one to two years.

Shrinkage

Shrinkage is caused by overdrying, and by the type of fabric being dried. To reduce the shrinking of cotton and rayon knits, it is recommended that, while the clothing is still damp, the user remove them from the dryer and lay them on a flat surface to air dry. Also, you might suggest to the user to set the heat setting on a lower setting. Before drying any synthetic clothing in a dryer, the user should read the label on the garment for the proper drying instructions, and for the proper heat selection. The user should read the use and care manual for the proper instructions when operating the dryer.

Stains

Greasy looking stains are caused by:
1. Fabric softener that is designed for use in the dryer.
2. Fabric softener that was undiluted when the clothes were washed.
3. Not enough detergent being used in the wash. This will cause the soil in the water to stick to the outer tub, and to return to the next load being washed.
4. Drying of clothing that already has a stain on it.

Brown stains on the clothing are caused by:
1. Fabric softener that is designed for use in the dryer.
2. Leaving the wash in the washer tub, after the cycle is completed.
3. Leaving the wet clothing in the dryer, for an extended amount of time.
4. Transmission seal leaking.

Static electricity

Static electricity in the clothes is caused by overdrying the clothes. To reduce this condition, you must instruct the user to reduce the amount of time the clothes are to be dried, and to use a lower temperature setting.

Static electricity in synthetics is normal. To reduce this condition, use a liquid fabric softener in the wash cycle or use a fabric softener that is designed for use in the dryer.

Wrinkling

Wrinkling of the clothes is caused by:
1. Not using the permanent press cycle which has a cooling down period, during which the heating element is not on.
2. Clothing not removed from the dryer, after it has completed the cycle.
3. Overloading the dryer. The clothes need the room to tumble freely without getting tangled.
4. The quality of the permanent press material in the garment might be poor.

AUTOMATIC DRYER MAINTENANCE

Automatic dryers and exhaust ducts must be cleaned periodically. The excess lint must be removed to prevent the possibility of a fire, and the possibility of the dryer not functioning properly. Most models have a service panel on the front of dryer, for accessibility. The outside of the cabinet should be wiped with a damp cloth. It is recommended that the components be inspected for wear and tear; if repairs are needed, they should be made as soon as possible.

REPAIR PROCEDURES

Each repair procedure is a complete inspection and repair process for a single dryer component.

Dryer timer

The typical complaints associated with the dryer timer are: 1. The dryer will not run at all. 2. The clothes are not drying. 3. The dryer will not stop at the end of the cycle. 4. The timer will not advance through the cycle.
1. *Verify the complaint* Verify the complaint by operating the dryer through its cycles. Before you change the timer, check the other components controlled by the timer.
2. *Check for external factors* You must check for external factors not associated with the appliance. Is the appliance installed properly? Does the appliance have the correct voltage?
3. *Disconnect the electricity* Before working on the dryer, disconnect the electricity. This can be done by pulling the plug from the electrical outlet. Or disconnect the electricity at the fuse panel or at the circuit breaker panel. Turn off the electricity.
4. *Remove the console panel to gain access* To gain access to the timer, remove the screws that secure the console to the top of the dryer (Fig. 11-3). On other models, to gain access, remove the back panel behind the console.
5. *Test the timer* There are two ways to check the timer:
   A. Connect a 120-volt fused service cord to the timer motor to see if the timer motor is operating (Fig. 11-4). Connect the ground wire of the 120-

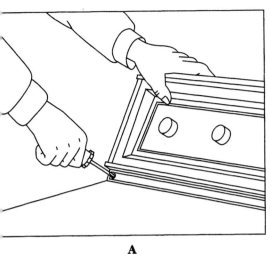

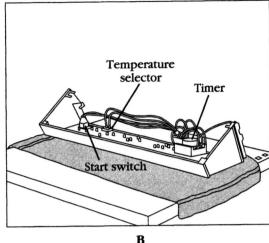

Temperature selector

Timer

Start switch

A   B

1-3 To gain access to the component, remove the screws. Rest the console panel on its face to gain access to the component.

volt fused service cord to the console ground. Be cautious when working with live wires. Avoid getting shocked.

The timer motor operates on 120 volts. If the motor does not operate, replace the timer. If the timer motor runs, but does not advance the cams, then the timer has internal defects, and it should be replaced.

B. Set the ohmmeter range to R × 100, disconnect the timer motor leads, and check for resistance (Fig. 11-5). The meter should read between 2000 and 3000 ohms. Next, test the timer switch contacts with the wiring diagram's configuration for the affected cycle. Place the meter probe on each terminal being tested, and turn the timer knob. If the switch contact is good, your meter will show continuity.

6. *Remove the timer* To remove the timer, pull the timer knob from the timer stem; then remove the timer mounting screws. Remove the wire lead terminals from the timer. Mark the wires as to their location on the timer. Some timers have a quick disconnect, instead of individual wires, which makes it easier to remove the timer wires (Fig. 11-6).

7. *Install a new timer* To install a new timer, just reverse the disassembly procedure, and reassemble. Replace the wires on the timer. Reinstall the console panel, and restore the electricity to the dryer. Test the dryer for proper operation.

Dryer motor

The typical complaints associated with motor failure are: 1. Fuse is blown, or the circuit breaker trips. 2. Motor will not start; only hums. 3. The dryer will not run.

1. *Verify the complaint* Verify the complaint by operating the dryer through its cycles. Listen carefully, and you will hear if there are any unusual noises, or if the circuit breaker trips.

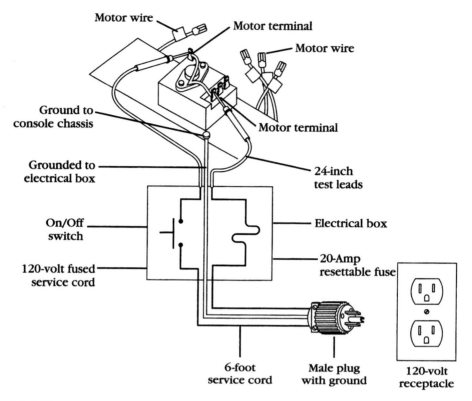

Motor wire

Motor terminal

Motor wire

Ground to console chassis

Motor terminal

Grounded to electrical box

24-inch test leads

On/Off switch

Electrical box

120-volt fused service cord

20-Amp resettable fuse

120-volt receptacle

6-foot service cord

Male plug with ground

11-4 The test cord attached to the timer motor leads.

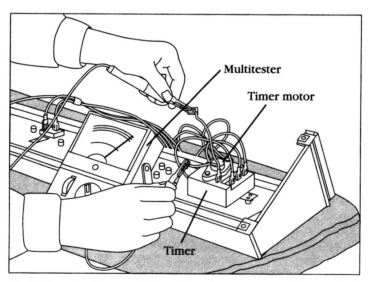

Multitester

Timer motor

Timer

11-5 Checking for continuity of the timer motor and the switch contacts.

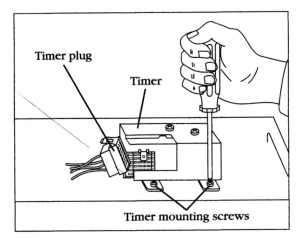

Timer plug

Timer

Timer mounting screws

11-6
Removing the timer dial. Then, remove the timer mounting screws, and lift the timer off the mounting bracket.

2. *Check for external factors* You must check for external factors not associated with the appliance. Is the appliance installed properly? Does the appliance have the correct voltage?

3. *Disconnect the electricity* Before working on the dryer, disconnect the electricity. This can be done by pulling the plug from the electrical outlet. Or disconnect the electricity at the fuse panel or at the circuit breaker panel. Turn off the electricity.

4. *Gain access to the motor* In order to gain access to the motor, the top must be raised (Fig. 11-7) by removing the screws from the lint screen slot. Insert a putty knife about two inches from each corner; then, disengage the retaining clips and lift the top. On some models, the top is held down with screws (see inset in Fig. 11-7). Remove the screws and lift the top. Now, the front panel of the dryer must be removed (Fig. 11-8). To remove the toe panel, insert a putty knife and disengage the retaining clip (Fig. 11-8A).

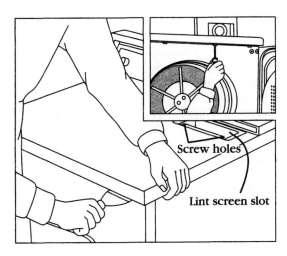

Screw holes

Lint screen slot

11-7
Inserting a putty knife to disengage the retaining clips that hold the top down. Don't forget to remove the two screws under the lint screen cover. On some models, remove the screws from underneath the top. See the insert.

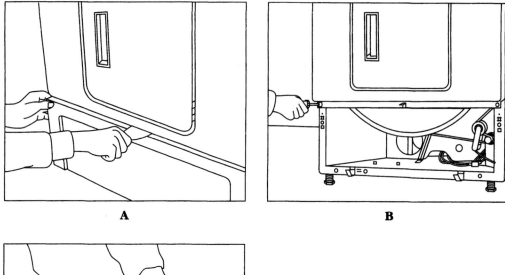

A

B

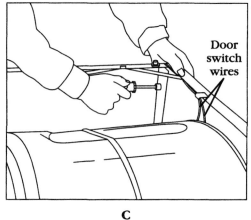

Door
switch
wires

C

11-8
Removing the toe panel.
Remove the screws from the
lower front panel. Remove the
screws from the upper front
panel. Remember to remove
the door switch wires.

On some models, the toe panel is held in place with two screws. In Figure 11-8B, remove the screws that hold the lower part of the front panel in place. Next, remove the screws that hold the upper part of the front panel in place (Fig. 11-8C), and then disconnect the door switch wires (label them). With the front panel out of the way, you can now disconnect the drive belt (Fig. 11-9). Push on the idler pulley to release the tension from the drive belt, and remove the belt from the motor pulley. Grab the drum and remove it from the cabinet (Fig. 11-10). On some models, you will have to remove the back panel because the drum comes out through the rear of the cabinet.

5. *Disconnect the motor wire leads* Disconnect the motor wire leads from wiring harness. Set the ohmmeter on R × 1. Figure 11-11 illustrates testing the motor windings and the centrifugal switch for continuity. When testing for resistance on the motor, test from the common wire lead to the run winding. Then, test for resistance from the common wire lead to the start winding. Next, test for resistance from the start winding to the run winding.

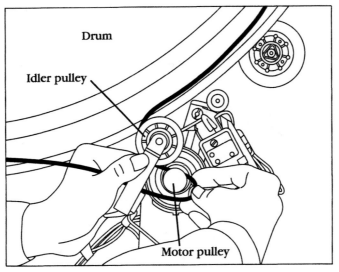

11-9 Pushing the idler pulley assembly toward the drive motor pulley to release the belt tension. Then, disconnect the drive belt from the idler pulley and drive motor pulley.

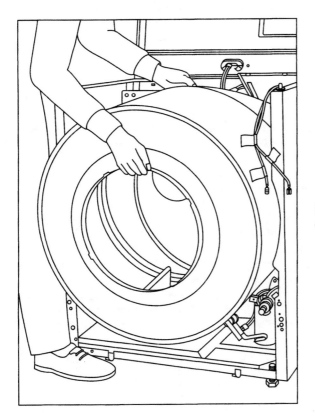

11-10
Removing the drum from the dryer cavity.

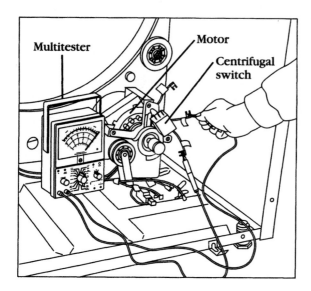

11-11
Testing the centrifugal switch and drive motor windings.

To test for a grounded winding in the motor, take the ohmmeter probes and test from each motor wire lead terminal to the motor housing. The ohmmeter will indicate continuity, if the windings are grounded. If the ohmmeter reading shows no resistance between the motor windings, then replace the motor. If the motor checks good, then check the timer.

6. *Remove the motor* To remove this type of motor, you must first disconnect the blower assembly by holding the motor shaft stationary, and then turning the blower wheel to remove it from the rear of the motor shaft (Fig. 11-12). Then, remove the spring clamps that hold the motor in the motor bracket (Fig. 11-13). On some models, the motor pulley must be removed (new motors come without the pulley attached), by loosening the allen set screw.

7. *Install the new motor* To install the new motor (Fig. 11-14), just reverse the disassembly procedure, and reassemble. Then, reassemble the dryer in the reverse order of its disassembly. Restore the electricity to the dryer and test the motor.

Drive belt

The typical complaints associated with belt failure are: 1. The drum will not turn. 2. Dryer motor spins freely. 3. Smells like something is burning.

1. *Verify the complaint* Verify the complaint by operating the dryer through its cycles. Listen carefully, and you will hear if there are any unusual noises. Then, with the door open, press the door switch and start the dryer. The drum should rotate.

2. *Check for external factors* You must check for external factors not associated with the appliance. Is the appliance installed properly? Does the appliance have the correct voltage? Check for foreign objects lodged between the drum and bulkhead, etc.

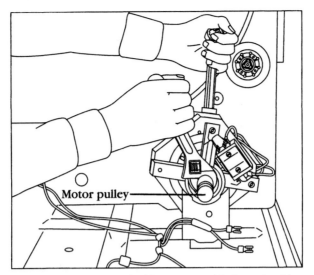

11-12 Disconnecting the blower assembly from the drive motor shaft.

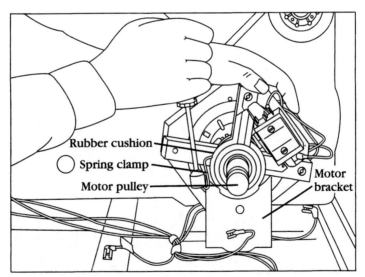

11-13 Holding the motor and remove the spring clamps.

3. *Disconnect the electricity* Before working on the dryer, disconnect the electricity. This can be done by pulling the plug from the electrical outlet. Or disconnect the electricity at the fuse panel or at the circuit breaker panel. Turn off the electricity.

4. *Gain access to the drive belt* To gain access to the drive belt, the top must be raised (Fig. 11-7) by removing the screws from the lint screen slot. Then, insert a putty knife about 2 inches from each corner, and disengage the

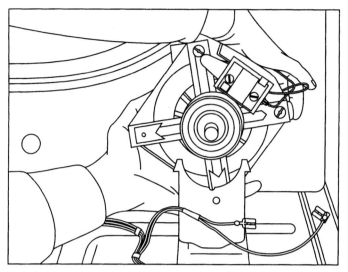

11-14 Placing the motor in the cradle and position it to fit into the slots properly.

retaining clips and lift the top. On some models, the top is held down with screws (see the inset in Figure 11-7). Remove the screws and lift the top. Now, the front panel of the dryer must be removed (Fig. 11-8). To remove the toe panel, insert a putty knife and disengage the retaining clip (Fig. 11-8A). In Figure 11-8B, remove the screws that hold the lower part of the front panel in place. Next, remove the screws that hold the upper part of the front panel in place (Fig. 11-8C), and then disconnect the door switch wires.

5. *Remove the drive belt* To remove the drive belt on this type of dryer, you can now disconnect the belt (Fig. 11-9). Push on the idler pulley to release the tension from the drive belt. Now, remove the belt from the motor pulley, and from around the drum (Fig. 11-15). If the drive belt is broken, just remove the belt.

6. *Install a new drive belt* To install the new drive belt, just reverse the disassembly procedure, and reassemble. Then, reassemble the dryer in the reverse order of its disassembly. Restore the electricity to the dryer and test.

Heater element

The typical complaints associated with heater failure are: 1. The dryer will not dry the clothes properly. 2. There is no heat at all when a heat cycle is selected.

1. *Verify the complaint* Verify the complaint by operating the dryer through its cycles. Then, open the dryer door, and place your hand inside to see if it is warm in the drum.

2. *Check for external factors* You must check for external factors not associated with the appliance. Is the appliance installed properly? Is the exhaust vent clogged? Check the voltage to the dryer. Note: the dryer motor runs on 120 volts, but the heater works on 240 volts.

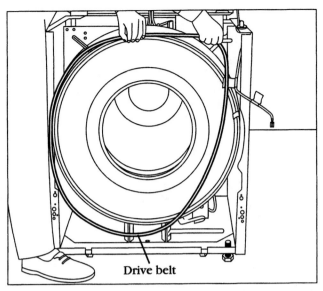

11-15 Removing the drive belt from around the drum.

3. *Disconnect the electricity* Before working on the dryer, disconnect the electricity. This can be done by pulling the plug from the electrical outlet. Or disconnect the electricity at the fuse panel or at the circuit breaker panel. Turn off the electricity.

4. *Gain access to the heater assembly* You must gain access to the heater assembly through the back. Remove the exhaust duct from the dryer. Then, remove the screws from the back panel (Fig. 11-16).

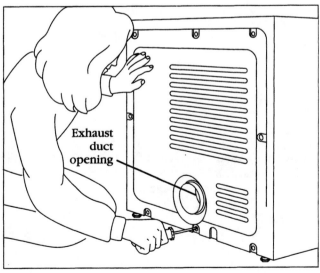

11-16 Removing the screws that hold the back panel in place.

5. *Test the heater element* To test the heater element, set the ohmmeter on R × 1. Remove the wires from the heater (Fig. 11-17), and test for continuity. If it is ok, check the thermostats.

6. *Remove the heater element* To remove the heater in this type of dryer, you must first remove the screw and the bracket that holds the heater box in place. It is located on top of the heater box. You can gain access through the back of the dryer, or lift the top and reach in near the back bulkhead. Then, slide the heater box up, and pull it away at the bottom (Fig. 11-18). With the heater box out of the dryer, remove the screw that holds the heater element in place (Fig. 11-19). Pull the heater element out of the heater box.

7. *Install new heater element* To install the new heater element, just reverse the disassembly procedure, and reassemble. Then, reinstall the back panel. Restore the electricity to the dryer and test it.

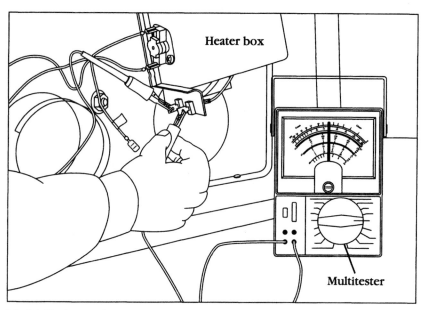

11-17 Testing the heater element for continuity.

Door switch

The typical complaints associated with switch failure are: 1. Dryer will not operate at all. 2. Dryer light not working when the door is open.

1. *Verify the complaint* Verify the complaint by operating the dryer through its cycles. Open the door to see if the light is working.

2. *Check for external factors* You must check for external factors not associated with the appliance. Is the appliance installed properly? Does the dryer have the correct voltage supply?

3. *Disconnect the electricity* Before working on the dryer, disconnect the electricity. This can be done by pulling the plug from the electrical outlet.

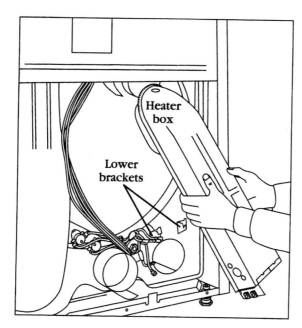

11-18
Removing the heater box.

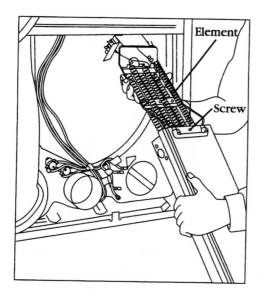

11-19
Removing the heater coil from the heater box.

Or disconnect the electricity at the fuse panel or at the circuit breaker panel. Turn off the electricity.

4. *Gain access to the door switch* You must gain access to the door switch. On this model, the top must be raised (Fig. 11-7), by removing the screws from the lint screen slot. Then, insert a putty knife about 2 inches from each corner, disengage the retaining clips, and lift the top. On some models the top is held down with screws (see the inset in Figure 11-7). Remove the screws and lift the top.

5. *Test the door switch* The door switch is located in one of the upper corners of the inside of the front panel (Fig. 11-20A). Once you have located the door switch, disconnect the wires from the terminals. Set the ohmmeter on R × 1, attach the probes of the ohmmeter to the terminals of the door switch, and then close the door. With the door closed, there should be continuity. With the door open, the ohmmeter should not show continuity. Some dryer models have a light inside the drum. This light circuit is also part of the door switch circuitry. There will be three terminals located on the door switch. Take your ohmmeter probe and place it on the common terminal of the door switch.

   To locate the common terminal, read the wiring diagram. It will indicate which terminal is the common, which is the light and is for the motor circuit. Take your probe and place it on the switch terminal (the light circuit), then close the dryer door. The reading should indicate no continuity. Now, open the dryer door, you should have a continuity reading on the meter.

6. *Remove the door switch* To remove the lever-type door switch (Fig. 11-20B), remove the screws that hold the switch in place. Now, lift the switch out from behind the front panel. To remove the cylindrical type of door switch (Fig. 11-20C), squeeze the retaining clips on the back side of the front panel, and pull out the switch. To remove the hinge type of door switch (Fig. 11-20D), the front panel must be removed. Next, remove the screws that hold the switch assembly in place.

7. *Install the new door switch* To install the new door switch, reverse the disassembly procedure, and reassemble. Reassemble the dryer, and install the wires on the new switch terminals. Plug in the dryer and test it.

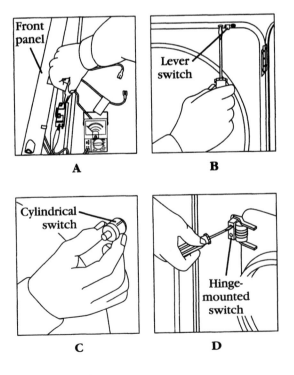

11-20
A. Testing the door switch for continuity. B. Removing the screws that secure the door switch. C. Removing a cylindrical door switch. D. Removing a hinge-mounted door switch.

Thermostat

The typical complaints associated with thermostat failure are: 1. The clothes are not drying. 2. The dryer will not shut off. 3. The dryer will not heat at all. 4. The drying temperature is too high. 5. Moisture retention (of fabrics) is unsatisfactory.

1. *Verify the complaint* Verify the complaint by operating the dryer through its cycles.
2. *Check for external factors* You must check for external factors not associated with the appliance. Is the appliance installed properly? Does the dryer have the correct voltage supply? Is the exhaust vent blocked?
3. *Disconnect the electric* Before working on the dryer, disconnect the electricity. This can be done by pulling the plug from the electrical outlet. Or disconnect the electricity at the fuse panel or at the circuit breaker panel. Turn off the electricity.
4. *Gain access to the thermostat* You must gain access to the thermostat through the back on this model. Remove the screws from the back panel (Fig. 11-16).
5. *Test the thermostat* To test the thermostat for continuity, remove the wires from the thermostat terminals. With a continuity tester, or an ohmmeter, test the thermostat for continuity (Fig. 11-21). Do this to all of the thermostats in the dryer. On some models, the thermostats might be located on the heater housing, behind the drum, or in the lint screen opening (Fig. 11-22). If they all check good, then reassemble the dryer and test for temperature operation. Take note, do not reinstall the back panel. Take a piece of paper and write down the temperature ratings of the thermostats. These ratings are printed on the thermostats (L140, L290, etc.). To test for temperature operation, you will need a voltmeter and a temperature tester.

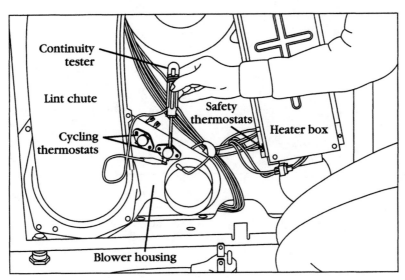

11-21 Using a continuity tester to check the thermostats.

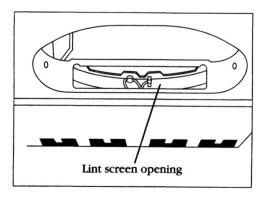

Lint screen opening

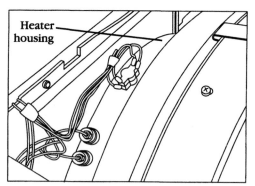

Heater housing

11-22
The location of thermostats on some other models.

First, set up the test instruments for testing the thermostatic operation. Take the temperature tester thermocouple lead and insert it in between the thermostat mounting ear (Fig. 11-23) and the plate against which it mounts. Then, connect the voltmeter probes across the thermostat terminals. Use alligator clips attached to the probe tips. This will allow you freedom of movement. Do not disconnect the wires from the thermostat. Set the voltmeter range to ac voltage, and the selector switch on 300 volts. Another thing to remember, if there is more than one thermostat (in parallel with other thermostats), the thermostats not under test must be electrically isolated. Review the wiring schematic to determine which wires to remove and which ones to isolate. With the test meters in place, you are now ready to remove the exhaust vent duct from the dryer. Seal off 75% of the exhaust opening from the dryer (Fig. 11-24). This will simulate a load of clothing in the dryer.

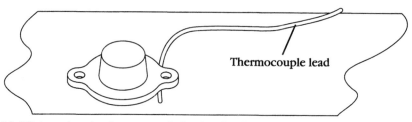

Thermocouple lead

11-23 Inserting the thermocouple lead under the thermostat ear.

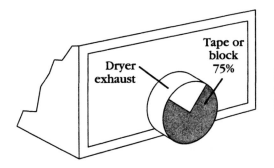

11-24
Blocking (or tape) 75% of the dryer exhaust to simulate a load.

Restore the electricity to the dryer. This test requires that the electricity be turned on for its duration. Always be cautious when working with live wires. Avoid getting shocked.

Set the controls on the dryer to operate at a high heat. Turn on the dryer and let it cycle. When the voltmeter is reading voltage, the thermostat has opened. When there is no voltage reading on the voltmeter, the thermostat is closed. As the dryer is cycling, via the thermostat, record the temperature. The thermostat should open at the preset temperature listed on the thermostat. Table 11-1 illustrates the types of thermostat with their opening and closing temperature settings. The temperature range of the thermostat should be within ±10% of the printed setting. If not, replace the thermostat.

Caution: turn off the electricity before replacing any parts of the dryer.

Table 11-1

Open	Close	Type	Open	Close	Type
260°	210°	L260	120°	110°	L120
270°	230°	L270	125°	115°	L125
290°	250°	L290	130°	115°	L130
300°	250°	L300	135°	120°	L135
310°	270°	L320	140°	130°	L140
340°	300°	L340	145°	125°	L145
120°	105°	LD120	150°	130°	L150
130°	115°	LD130	155°	135°	L155
140°	120°	LD140	160°	120°	L160
145°	125°	LD145	170°	130°	L170
155°	135°	LD155	180°	170°	L180
170°	150°	LD170	190°	150°	L190
200°	160°	LD200	200°	160°	L200
225°	185°	LD225	205°	165°	L205
240°	195°	LD240	225°	185°	L225
270°	225°	LD270	240°	200°	L240
290°	250°	LD290	250°	210°	L250

6. *Replace the thermostat* Remove the screws that hold the thermostat in place. Replace the thermostat with an exact replacement of the same temperature rating. Reconnect the wires to their correct terminal positions. Then, reverse the disassembly procedure to reassemble the dryer; and test the thermostat.

7. *Test the new thermostats* To test the new thermostats, repeat step 5.

Start switch

The typical complaint associated with the start switch is that the dryer will not start.

1. *Verify the complaint* Verify the complaint by operating the dryer. Before you change the start switch, check the other components that are in the start circuit.

2. *Check for external factors* You must check for external factors not associated with the appliance. Is the appliance installed properly? Does the appliance have the correct voltage?

3. *Disconnect the electricity* Before working on the dryer, disconnect the electricity. This can be done by pulling the plug from the electrical outlet. Or disconnect the electricity at the fuse panel or at the circuit breaker panel. Turn off the electricity.

4. *Remove the console panel to gain access* Begin with removing the console panel, to gain access to the start switch. With this type of dryer, remove the screws from the console (Fig. 11-3A). On some models, the console will be able to lay flat (Fig. 11-3B).

5. *Test the start switch* Remove the wires from the start switch terminals. Set the ohmmeter on R × 1. Begin testing the start switch for continuity by placing one probe of the ohmmeter on the common terminal (C) of the switch, then connect the other probe to the normally open (NO) terminal. There should be no continuity. With the ohmmeter probes attached to the switch, press the start switch button. You should have continuity (Fig. 11-25). If the switch fails the test, replace it.

6. *Remove the start switch* To remove the start switch, pull the knob off the switch stem, remove the start switch mounting screws, and remove the switch.

7. *Install a new start switch* To install a new start switch, just reverse the disassembly procedure, and reassemble. Replace the wires on the switch. Reinstall the console panel, and restore the electricity to the dryer. Test the dryer operation.

Drum roller

The typical complaints associated with the drum roller are: 1. Dryer very noisy when operating. 2. Dryer has a burning smell.

1. *Verify the complaint* Verify the complaint by operating the dryer through its cycles.

2. *Check for external factors* You must check for external factors not associated with the appliance. Is the appliance installed properly? Does the dryer have the correct voltage supply?

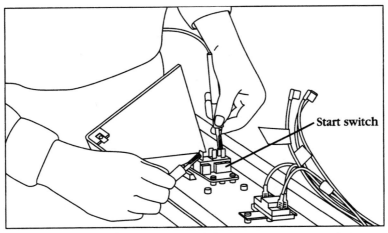

11-25 Testing the continuity of the switch contacts on a start switch.

3. *Disconnect the electricity* Before working on the dryer, disconnect the electricity. This can be done by pulling the plug from the electrical outlet. Or disconnect the electricity at the fuse panel or at the circuit breaker panel. Turn off the electricity.

4. *Gain access to the drum roller* To gain access to the drum roller, the top must be raised (Fig. 11-7) by removing the screws from the lint screen slot. Then, insert a putty knife about two inches from each corner to disengage the retaining clips, and lift the top. On some models, the top is held down with screws (see the inset in Fig. 11-7). Remove the screws and lift the top. Now, the front panel of the dryer must be removed (Fig. 11-8). To remove the toe panel, insert a putty knife and disengage the retaining clip (Fig. 11-8A). As in Fig. 11-8B, remove the screws that hold the lower part of the front panel in place, then remove the screws that hold the upper part of the front panel in place (Fig. 11-8C). Disconnect the door switch wires.

5. *Remove the drive belt and drum* To remove the drive belt and the drum on this type of dryer, you must first disconnect the belt (Fig. 11-9). Push on the idler pulley to release the tension from the drive belt. Remove the belt from the motor pulley and from around the drum (Fig. 11-15). Grab hold of the drum and remove it from the cabinet (Fig. 11-10).

6. *Remove the drum roller* Once the drum has been removed, the drum rollers, which are located on the rear bulkhead of this model (Fig. 11-26), can then be removed. Take a pair of needle-nose pliers and remove the tri-ring off the shaft. Then, slide the roller off the shaft.

7. *Install a new drum roller* To install the new roller, clean the shaft and lubricate it with a small amount of oil. Slide the new drum roller onto the shaft, and reinstall the tri-ring. To reassemble the dryer, just reverse the disassembly procedure, and reassemble. Test the dryer for proper operation.

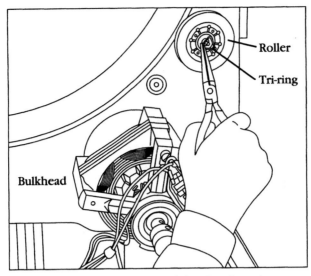

11-26 You must remove the tri-ring first, in order to remove the drum roller.

Idler pulley

The typical complaints associated with the idler pulley are: 1. When dryer is running, you will hear a squealing noise. 2. The dryer belt burns, and might snap. 3. The dryer drum will not turn.

1. *Verify the complaint* Verify the complaint by operating the dryer through its cycles.
2. *Check for external factors* You must check for external factors not associated with the appliance. Is the appliance installed properly? Does the dryer have the correct voltage supply?
3. *Disconnect the electricity* Before working on the dryer, disconnect the electricity. This can be done by pulling the plug from the electrical outlet. Or disconnect the electricity at the fuse panel, or at the circuit breaker panel. Turn off the electricity.
4. *Gain access to the idler pulley* To gain access to the idler pulley, in this type of dryer, remove the toe panel, insert a putty knife, and disengage the retaining clip (Fig. 11-8A).
5. *Remove the idler pulley* To remove the idler pulley, you must first remove the drive belt. To disconnect the belt on this model (Fig. 11-9), push on the idler pulley to release the tension from the drive belt. Now, remove the belt from the motor pulley, and from around the idler pulley. If the drive belt is broken, just remove the belt. Next, lift the idler pulley up and out of the dryer (Fig. 11-27). This type of idler pulley is replaced as a complete assembly. Another type of idler pulley (Fig. 11-28A) has a tension spring. To remove this type of idler, you must first remove the tension spring (Fig. 11-28B), and then lift the idler assembly out of the dryer. Inspect the pulley for wear. If it is worn, remove the screw from the axle, and pull the axle out (Fig. 11-28C). On this type of idler pulley, only replace the worn out part.

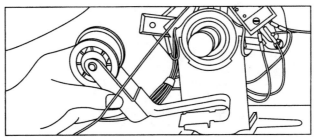

11-27 After removing the drive belt, lift the idler pulley off the base.

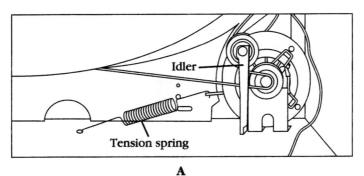

Idler

Tension spring

A

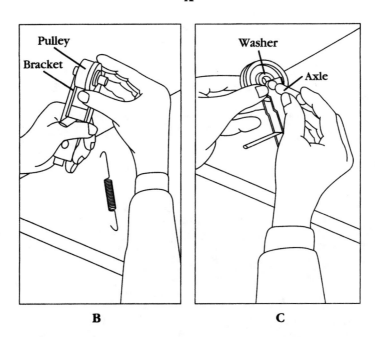

Pulley

Bracket

Washer

Axle

B   **C**

11-28 A. This type of idler pulley uses a spring to hold tension against the drive belt. B. Remove the spring and lift the idler assembly out of the dryer for inspection. C. Remove the axle to replace the pulley.

6. *Install the new idler pulley* To install the repaired idler pulley, just reverse the disassembly procedure, and reassemble it. Reassemble the dryer in the reverse order of its disassembly and test it for proper operation.

Temperature selector switch

The typical complaints associated with the temperature selector switch are: 1. Inability to select a certain temperature setting. 2. No heat.

1. *Verify the complaint* Verify the complaint by operating the dryer through its cycles. Before you change the temperature selector switch, check the other components controlled by this switch.

2. *Check for external factors* You must check for external factors not associated with the appliance. Is the appliance installed properly? Does the appliance have the correct voltage?

3. *Disconnect the electricity* Before working on the dryer, disconnect the electricity to the dryer. This can be done by pulling the plug from the electrical outlet. Or disconnect the electricity at the fuse panel or at the circuit breaker panel. Turn off the electricity.

4. *Gain access to the temperature selector switch* To access the temperature selector switch, remove the console panel. On this type of dryer, remove the screws in the console (Fig. 11-3A). On some models, the console will be able to lay flat (Fig. 11-3B). On other models, your access is through the rear panel on the console.

5. *Test the temperature selector switch* To test the temperature selector switch, locate the selector switch circuit on the wiring diagram. Identify the terminals that are regulating the temperature setting to be tested. Set the ohmmeter on the R × 1 scale. Next, place the ohmmeter probes on those terminals. Then, select that temperature setting by either rotating the dial or by depressing the proper button on the switch (Fig. 11-29). If the switch

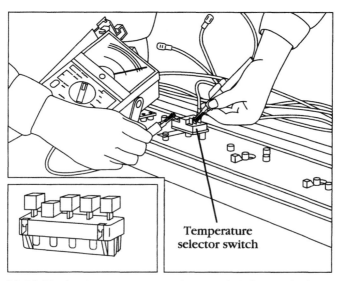

11-29 Testing the temperature selector switch for continuity.

contacts are good, your meter will show continuity. Test all of the remaining temperature settings on the temperature selector switch. Remember to check the wiring diagram for the correct switch contact terminals (those that correspond to the setting that you are testing).

6. *Remove the temperature selector switch* To remove the temperature selector switch, remove the screws that hold the switch to the console base, and remove the switch.

7. *Install the new temperature selector switch* To install the new temperature selector switch, just reverse the disassembly procedure and reassemble it. Then, reattach the wires to the switch terminals according to the wiring diagram. Reassemble the console panel. Be sure that, when you are reassembling the console panel, the wires do not become pinched between the console panel and the top of the dryer.

DIAGNOSTIC CHARTS

The following diagnostic flow charts will help you to pinpoint the likely causes of the dryer problems (Figs. 11-30, 11-31, and 11-32).

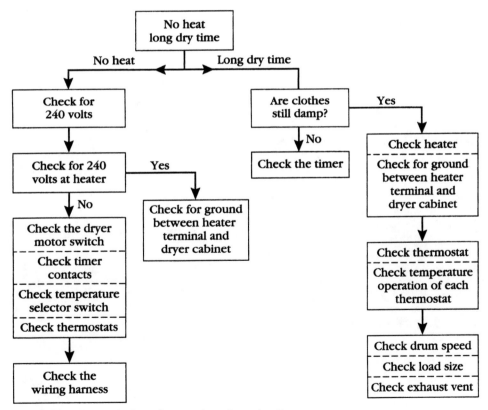

11-30 The diagnostic flow chart: no heat, long dry-time.

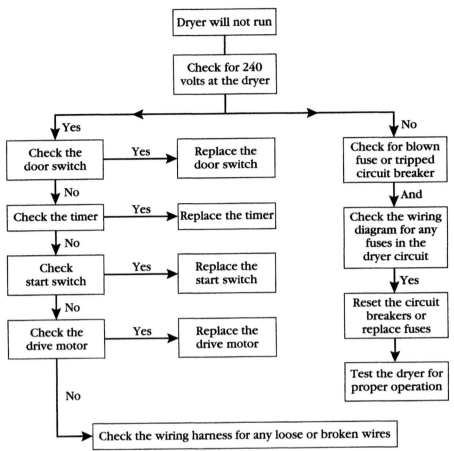

11-31 The diagnostic flow chart: dryer will not run.

The wiring diagram in this chapter is only an example. You must refer to the wiring diagrams on the dryer that you are servicing. Figure 11-33 depicts an actual wiring schematic diagram and an actual pictorial electrical wiring diagram. Figure 11-34 depicts an actual ladder diagram. A ladder diagram is simpler and it is usually easier to read.

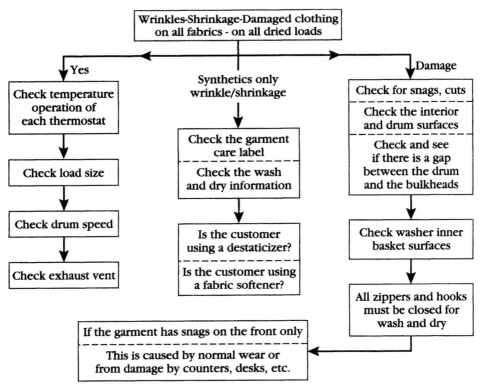

Wrinkles-Shrinkage-Damaged clothing
on all fabrics - on all dried loads

Yes

Check temperature
operation of
each thermostat

Check load size

Check drum speed

Check exhaust vent

Synthetics only
wrinkle/shrinkage

Check the garment
care label

Check the wash
and dry information

Is the customer
using a destaticizer?

Is the customer using
a fabric softener?

Damage

Check for snags, cuts

Check the interior
and drum surfaces

Check and see
if there is a gap
between the drum
and the bulkheads

Check washer inner
basket surfaces

All zippers and hooks
must be closed for
wash and dry

If the garment has snags on the front only

This is caused by normal wear or
from damage by counters, desks, etc.

11-32 The diagnostic flow chart: wrinkles, shrinkage, and damaged clothing with all fabrics on all dried loads.

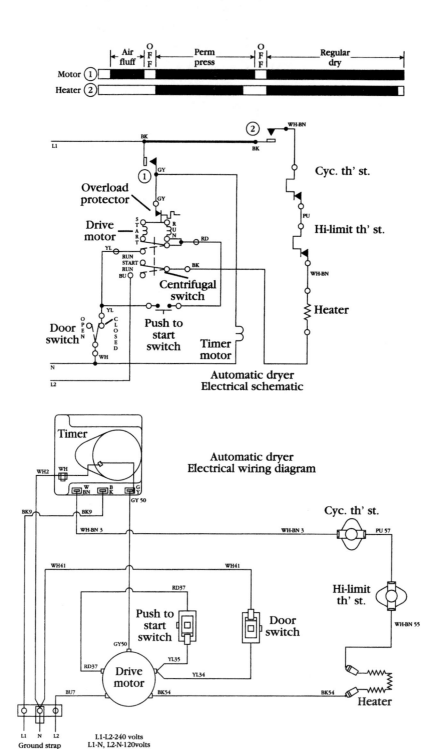

11-33 The automatic dryer wiring diagram.

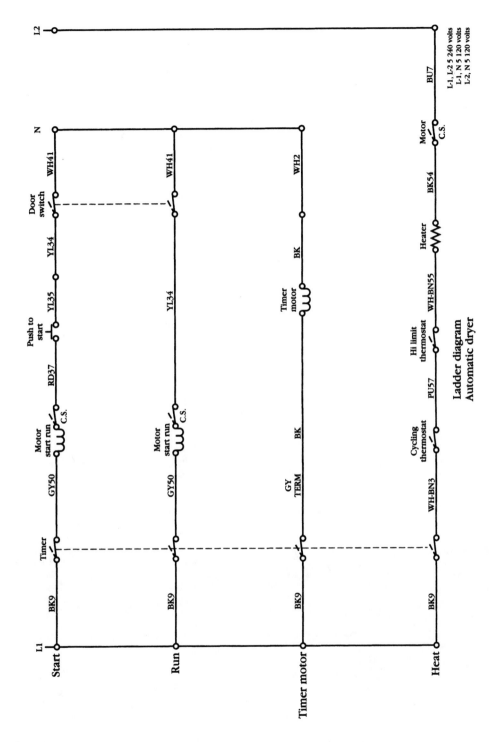

Ladder diagram
Automatic dryer

L-1, L-2 5 240 volts
L-1, N 5 120 volts
L-2, N 5 120 volts

11-34 The automatic dryer ladder diagram.

12
Electric ranges and ovens

Electric ranges, cooktops, and ovens are available in a variety of styles. Electric ranges and ovens might seem to be complicated, but they are not. The more you know about the electrical and mechanical operation of the product, the easier it will be to solve their problems. An electric range, cooktop, or oven operates on 240 volts for the heating elements, and 120 volts for the accessories (clock, lights, etc.). Most repairs are electrical in nature. This chapter will provide the basics needed to diagnose and repair these appliances. Figures 12-1 and 12-2 identify where components are located within the range, and these illustrations are used as examples only. The actual construction and features might vary, depending on what brand and model you are servicing. This chapter does not include any repairs for microwave ovens or ranges with a microwave feature.

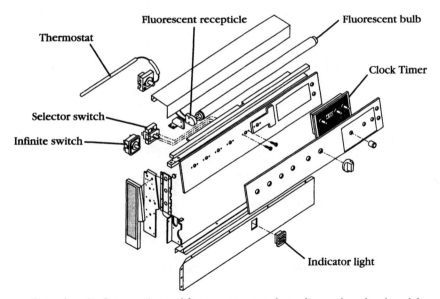

Example only: Construction and features may vary depending on brand and model

12-1 An example view of a range backguard. The construction and features might vary, depending on the brand and model.

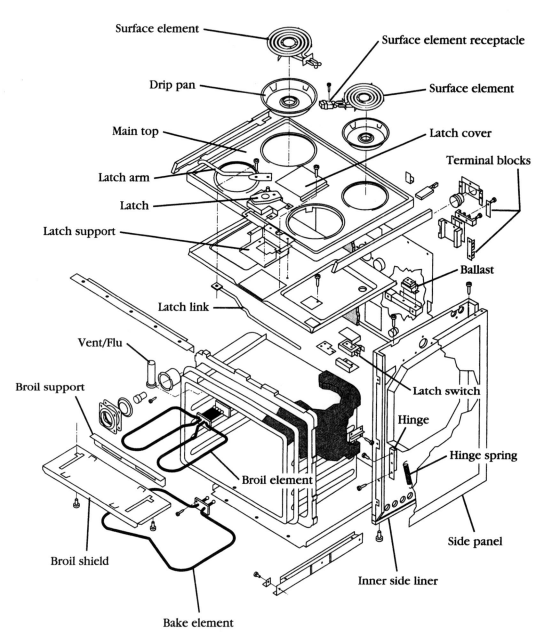

Surface element

Surface element receptacle

Drip pan

Surface element

Main top

Latch cover

Terminal blocks

Latch arm

Latch

Latch support

Ballast

Latch link

Vent/Flu

Broil support

Latch switch

Hinge

Hinge spring

Broil element

Bake element

Broil shield

Inner side liner

Side panel

Example only: Construction and features may vary depending on brand and model

12-2 An example view of a range chassis. The construction and features might vary, depending on the brand and model.

PRINCIPLES OF OPERATION

The surface units are very simple to operate. After placing a pot on the surface element, you then turn the surface unit switch to the desired setting. When the surface unit switch is on, current flows from the wall receptacle, through the wiring and the surface unit switch, and then through the heating element. To properly cook on the surface unit, the pot or pan must lay flat across the heating element (Fig. 12-3), making contact with the entire cooking surface. If not, the food will not cook evenly, and there will be greater heat losses (resulting in wasted electricity).

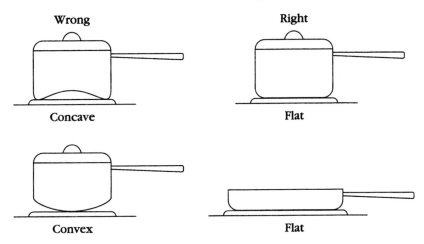

12-3 The cookware must make full contact with the surface unit.

To bake in the oven, place the food on the oven rack, and close the oven door. The oven selector switch is set to bake, and the oven temperature control is set on the desired temperature. Current flows from the wall receptacle, through the wiring and the selector switch, through the thermostat, and on to the bake element. When the selected temperature in the oven is reached, the bake element will go off. When the temperature in the oven decreases, the thermostat then reactivates the bake element. On some models, both the bake and broil elements operate at the same time to first preheat the oven cavity.

The broiling operation in the range is accomplished by placing the food on the top oven rack, and partially closing the oven door. When closing the door, there is usually a stop in the hinge that allows the door to remain open a certain amount; this is done so that the broiler element will not cycle on temperature. The broiler element stays on for continuous operation, until the user turns off the controls. When broiling, current will flow from the wall receptacle, through the wiring and the selector switch, through the thermostat, and then to the broil element.

Self-cleaning ranges and ovens differ (from manufacturer to manufacturer, and from model to model) in how the self-cleaning feature of the heating elements operates.

Pyrolytic cleaning is the true self-clean system. It uses high heat, during a special 1- to 3-hour cycle, to decompose food soil and grease. During the cycle, which is clock-controlled, the oven door is latched and locked. It cannot be opened until the oven cools down. All of the oven walls, racks, and the door (except for a small area outside the door gasket) are completely cleaned. After cleaning, you might find a small bit of white ash, which can be easily wiped out.

Catalytic, or continuous cleaning, uses a special porous coating on the oven walls that partially absorbs and disperses the soil. This process occurs during normal baking, and keeps the oven presentably clean; but the racks and door parts must be cleaned by hand. Some manufacturers recommend occasionally operating an empty oven at 500 degrees Fahrenheit to remove any build-up of soil. This special oven coating cannot be cleaned with soap, detergent, or commercial oven cleaners without causing permanent damage.

For example, on some models, you simply close and latch the oven door, and then set the controls to clean. When the self-clean cycle is going to be used, the consumer must read the use and care manual in order to set the controls properly. If the controls are not set according to the manufacturer's recommendations, the oven will not be properly cleaned.

Next, you set the time clock for the proper time when you want the clean cycle to begin. The clock in a self-clean oven has two functions: First, to control when the oven will operate in the timed-bake cycle; and second, for control during the clean cycle. The bake element will come on, and the temperature will begin to rise. This will take approximately 60 minutes. The reason for the slow temperature rise is to prevent damage to the oven cavity and the door.

When the temperature is above 550 degrees Fahrenheit, the oven lock light will come on, indicating that the latch on the oven door cannot be opened until the oven cavity cools below that temperature. When the oven cavity temperature stabilizes (between 840 and 920 degrees Fahrenheit), the cleaning process begins. This cleaning process requires approximately two to three hours (see Table 12-1 on page 299). Some models have a cooling fan in the circuit to aid in keeping the exterior temperature low. Self-clean ovens have a nonelectric, "catalytic" smoke eliminator in the vent to consume the smoke from the soil load. The catalytic smoke eliminator will begin to operate when the oven cavity temperature is between 300 and 400 degrees Fahrenheit. The types of oven door-lock mechanisms used on self-cleaning models are:

1. *Manual door lock* With this type of system, the user manually moves the oven door latch assembly to lock the door (see Fig. 12-4).
2. *Electromechanical door lock* With this type of system, the user will either press a lock switch button (located on the control panel) in order to move the latch handle; or, alternatively, move the latch handle to lock or unlock the oven door. This action will activate a solenoid coil, allowing the door latch mechanism and linkage to operate (Fig. 12-5).
3. *Electric door lock* With this type of system, the user sets the cleaning controls, and the door lock assembly will operate, through the control of an electric motor, to activate the locking mechanism. This system is similar to the electromechanical door lock; however, instead of a latch solenoid, an electric motor is used.

Manual latch system

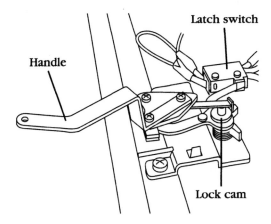

12-4
The latch solenoid system for the self-cleaning range oven is used on some models.

SAFETY FIRST

Any person who cannot use basic tools, or follow written instructions, should *not* attempt to install, maintain, or repair any electric range/oven/cooktop. Any improper installation, preventive maintenance, or repairs could create a risk of personal injury or property damage.

If you do not fully understand the installation, preventive maintenance, or repair procedures in this chapter, or if you doubt your ability to complete the task on your appliance, then please call your service manager.

The following precautions should also be followed:

1. Never use a range to heat the home, it simply wasn't designed for that purpose.
2. Keep the cooking area clean from spills and grease.
3. Do not use flammable liquids near a cooking appliance.
4. When repairing the range, always use the proper tools.
5. Always reconnect the ground wire to the range after repairs have been made.
6. Never use aluminum foil to line drip bowls; it could cause an electrical shock or become a fire hazard.

Before continuing, take a moment to refresh your memory of the safety procedures in Chapter 2.

ELECTRIC RANGES, OVENS, AND COOKTOPS IN GENERAL

Much of the troubleshooting information in this chapter covers electric range/oven/cooktops in general, rather than specific models, in order to present a broad overview of service techniques. The pictures and illustrations that are used in this chapter are for demonstration purposes to clarifying the description of how to service these appliances. They in no way reflect on a particular brand's reliability. Please note that *eye-level ranges* are also referred to as *high-low ranges* or *tri-level ranges*.

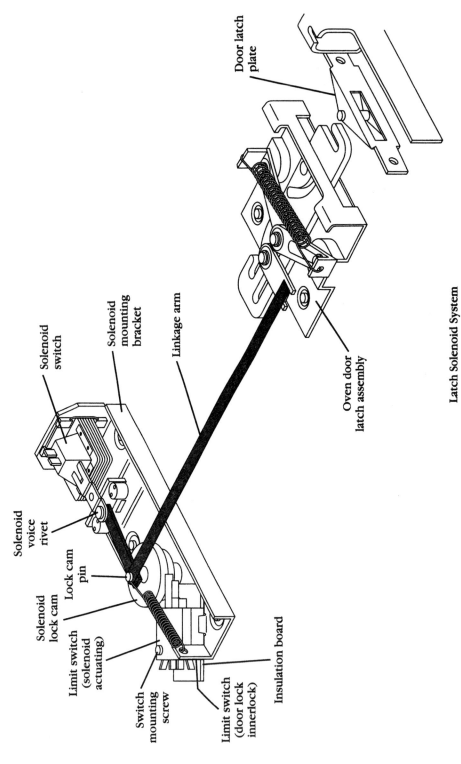

Solenoid switch

Solenoid mounting bracket

Linkage arm

Door latch plate

Solenoid voice rivet

Lock cam pin

Solenoid lock cam

Limit switch (solenoid actuating)

Switch mounting screw

Limit switch (door lock innerlock)

Insulation board

Oven door latch assembly

Latch Solenoid System

12-5 Turning off the electricity before servicing the appliance.

Location and installation of electric range, oven, and cooktop

Locate the range, oven, or cooktop where it will be well lighted. The range must be level for proper baking and cooking results. The range might be installed adjacent to the left and/or right base cabinets, and against a rear vertical wall (for the "anti-tip" cleat). A wall oven must be installed on a supporting surface that is strong enough to support the weight of the oven and its contents, while remaining level from side to side, and from front to rear. A cooktop must be installed on a flat surface, supported by the countertop, and it should be level. The proper installation instructions for your model are included. These instructions will assist you with the installation requirements (dimensions, electrical requirements, cutout dimensions, venting, etc.) needed to complete the installation, according to the manufacturers specifications.

STEP-BY-STEP TROUBLESHOOTING BY SYMPTOM DIAGNOSIS

When servicing an appliance, don't overlook the simple things that might be causing the problem. Step-by-step troubleshooting, by symptom diagnosis, is based on diagnosing malfunctions, with their possible causes arranged into categories relating to the operation of the range. This section is intended only to serve as a checklist, to aid you in diagnosing a problem. Look at the symptom that best describes the problem that you are experiencing with the range/oven/cooktop, then proceed to correct the problem.

Oven will not heat

1. Are the oven controls set properly?
2. If the entire range is inoperative; check for voltage, and check the circuit breakers or fuses.
3. Test oven thermostat switch contacts for continuity.
4. Test the heating elements for continuity, and for a short.
5. Check for loose or broken wiring connections.
6. Test the other components in circuits which operate the heating elements.
7. Test the oven selector switch.

Oven temperature is not accurate

1. Test the oven thermostat for accuracy, by using an oven temperature tester.
2. Check to be sure that the oven door is closed for baking.
3. Is the oven vent blocked with aluminum foil?

Oven will not cycle off

1. Test the oven thermostat switch contacts for continuity.
2. Test the oven selector switch contacts for continuity.
3. Test the time clock switch contacts for continuity.
4. Test the oven cycling relay.

Broil element not heating

1. Are the broil controls set properly?
2. If the entire range is inoperative; check for voltage, and check circuit breakers or fuses.
3. Test the oven thermostat switch contacts for continuity.
4. Test the heating element for continuity, and for a short.
5. Check for loose or broken connections.
6. Test other components in circuits which operate the heating elements.
7. Does the user have the door closed completely?
8. Check oven selector switch.
9. Check to be sure that the oven door is open for broiling.

Surface unit will not operate

1. Is the right surface unit switch turned on?
2. Check for loose or broken wiring connections.
3. Test the surface element for continuity, and for a short.
4. If the entire range is inoperative; check for voltage, and check circuit breakers or fuses.
5. Test the surface unit switch for continuity.

Heat escaping from oven door

1. Is the oven door aligned properly with the range body?
2. Check oven door gasket.
3. Check the oven door hinge for damage.
4. Check the oven door spring. Is it broken?
5. Is the consumer closing the door completely for baking?

Cooking performance

If there are no mechanical problems with the oven's operation, but it will not bake the food properly or if the food is partially cooked, etc.; your next step will be to look at the symptom that best describes the problem that you are experiencing with the oven. Then, correct the problem. If necessary, instruct the user how to get better results from the oven. This information is covered in the use and care manual.

Satisfactory baking results To have satisfactory baking results, the following conditions must exist:

1. Proper oven door seal.
2. The oven vent not blocked off.
3. The oven thermostat must be calibrated properly.
4. Food preparation must be done correctly.
5. The proper type of cookware is used.
6. Follow the manufacturer's recommended cooking instructions.

Satisfactory broiling results To have satisfactory broiling results, the following conditions must exist:

1. On most models, the oven door must be open partially, to ensure that the broiling element will not cycle on temperature.
2. Food preparation must be done correctly.
3. The proper type of cookware is used.
4. Follow the manufacturer's recommended broiling instructions.

Satisfactory surface cooking results To have satisfactory surface cooking results, the following conditions must exist:
1. The reflector pans (bowls) must be used under all of the surface units.
2. Use cookware with flat bottoms, that will correctly contact the heating element.
3. Use cookware large enough to entirely cover the surface units.
4. Food preparation must be done correctly.
5. Follow the manufacturer's recommended cooking instructions.

Surface cookware

One of the keys to successful cooking is the use of proper cookware. For efficiency and best results, use pans that:
- Have tight-fitting lids
- Have lightweight handles that do not tilt the pans

Listed is a brief review of the various types of cookware that is available. The different characteristics of the cookware will aid you when diagnosing cooking complaints.
1. Aluminum is a metal that spreads heat quickly and evenly and responds to temperature changes. This cookware is best for frying, braising and pot roasts. The inside of an aluminum pan might be a natural finish or a nonstick coating.
2. Cast iron is slow to change temperature and holds heat. This makes good cookware for browning, frying, stewing and other cooktop cooking. Cast iron cookware is also available with an enamel finish.
3. Copper is excellent for gourmet cooking, wine sauces and egg cookery and it is quick to change temperature.
4. Glass cookware is slow to change temperature. This type of cookware works best for a long period, on a low heat cooking with a liquid.
5. Porcelain enamel is long lasting and it is used for cooking soups and other liquids.
6. Stainless steel is exceptionally strong. It is used for frying, sauces, soups, vegetables, and egg cooking.

Oven cookware

Sometimes a technician must educate the consumer on how to be a successful cook. The use of the correct type of cookware is very important. Listed below are some guidelines for choosing cookware.
1. Always use the correct size pan, as given in the recipe.
2. When baking foods, use flat-bottomed pans to keep the food level.
3. Aluminum oven cookware that is shiny (not cast) produces delicate browning, tender crusts, and reduces the spattering of roasts. This type of

cookware is best for cakes, muffins, some quick breads, cookies and roasting.

4. Pottery, ceramic, cast metal, and ovenproof glass cookware gives food a deep, crusty brown surface. When this type of cookware is used, the oven temperature should be reduced by 25 degrees Fahrenheit.

5. The use of dull, or darkened, cookware is suitable for pies and other foods baked in pastry shells.

6. Shallow-sided pans and flat baking sheets are best for cookies and biscuits, where top and side browning is wanted.

Food preparation

The preparation of the food is the other key to successful cooking. If you suspect a baking problem, ask the cook to see the cooked food. If needed, ask the cook to prepare some food for baking, and watch how the food is prepared. The following descriptions will assist you in determining the problem.

Flat cake The cake comes out of the oven and you notice that the cake is flat, or that it has no volume to it, the reasons are:

1. There might be too much liquid, or not enough liquid.
2. During preparation, the mix was overbeaten or underbeaten.
3. The pan used to make the cake in was too large.
4. The cook did not adjust the oven racks properly, or placed the pan on the wrong rack.
5. The oven temperature too high, or too low.
6. The mix was stored improperly and/or was exposed to high humidity.
7. The cook might have forgotten to use eggs.

Cake has fallen The cake comes out of the oven and you notice that the cake has fallen, or the center has a dip in it, the reasons are:

1. The cake is underbaked.
2. There might be too much liquid, or not enough liquid.
3. The cook tested the cake too soon.
4. The cake pan was too small.
5. The cook moved the cake before it was completely baked.

Sticky crust The cake comes out of the oven, and you notice that the cake has a sticky top crust. The reasons are:

1. The cake was underbaked.
2. The cook stored the cake in a sealed container when the cake was still warm.
3. The humidity in the air is high.
4. Too much liquid in the mix.
5. The cook might have substituted sweet fruit juices, for other liquids.

Holes in cake The cake comes out of the oven, and you notice that the cake has holes or tunnels in it. The reasons are:

1. Baking temperature selected was too high.
2. The cook did not adjust the oven racks properly, or placed the pan on the wrong rack.

3. During preparation, the mix was overbeaten or underbeaten.
4. Very large air bubbles became trapped in the batter.
5. The cook might have used the wrong type of pan, which might have caused uneven temperature conductance.

Cake shrinkage The cake comes out of the oven, and you notice that the cake shrank or pulled away from the sides of the pan. The reasons are:
1. The cake was overbaked.
2. The cake pan was too close to the oven wall, or too close to other pans.
3. Extreme overbeating of the mixture.

Cake peaked in the center The cake comes out of the oven, and you notice that the center of the cake has a peak in it. The reasons are:
1. The oven temperature selected was too high.
2. The pan was too small.
3. The cook might have used the wrong type of pan, which might have caused uneven temperature conductance.
4. There might not have been enough liquid used in the mixture.
5. The cake pan was too close to the oven wall, or too close to other pans.
6. Extreme overbeating of the mixture.

Crust not brown The cake comes out of the oven, and you notice that the crust on top of the cake is not brown. The reasons are:
1. Oven door opened too many times during baking.
2. Too much liquid in the mix.
3. Oven temperature too low when baking.
4. The cake pan was too deep.
5. Extreme overbeating of the mixture.

Cake too dry The cake comes out of the oven, and you notice that the cake is too dry, and that it falls apart, the reasons are:
1. The cake was overbaked.
2. Not enough liquid in the mix.
3. The cook might have forgotten to use eggs.

Cookies brown rapidly The cookies come out of the oven, and you notice that the cookies are browned more at the sides and/or the end of the cookie sheet, the reasons are:
1. The cook might have used the wrong size cookie sheet, which might have caused uneven temperature conductance.
2. Cookie sheets with sides will cause rapid browning at the edges. Are the cookies placed at least one inch from the sides?

Cookies brown slowly The cookies come out of the oven, and you notice that the cookies browned slowly. The reasons are:
1. The heat might be leaking out of the oven. Check the oven door for heat leakage.
2. The racks are uneven. Check the oven racks. Are they level when cold and when hot?

Cookies are dark on the bottom   The cookies come out of the oven, and you notice that the cookies are dark on the bottom. The reasons are:
1. The oven racks were not adjusted properly, or the cookie sheet was placed on the wrong rack.
2. More than one cookie sheet was placed into the oven.

Cooking meats

The cook complains that the meats are not cooking properly; ask to see some of the food. This section is intended only to serve as a checklist, to aid you in diagnosing a problem. Look at the symptom that best describes the problem that the cook is experiencing with the meats, then correct the problem.

The meat burns on the bottom   The possible reasons are:
1. Check the size of the pan used to cook the meat. There should be at least 1 or 2 inches of oven rack visible around the pan when it is placed on the oven rack so that the heat will heat the oven cavity evenly.
2. Was the oven preheated?
3. Was the door opened and closed frequently?
4. Was the meat elevated off the bottom of the pan?

Meats are undercooked   The possible reasons are:
1. Check the oven temperature. Is it calibrated correctly?
2. Check the size of the pan used to cook the meat. There should be at least 1 or 2 inches of oven rack visible around the pan, when it is placed on the oven rack so that the heat will heat the oven cavity evenly.
3. Was the type of pan used too deep? Was the pan covered?

Roasting of meats take too long   The possible reasons are:
1. Check the oven temperature. Is it calibrated correctly?
2. Check for inadequate ventilation.
3. Check for the improper use of aluminum foil.

RANGE, OVEN, AND COOKTOP MAINTENANCE

The range, oven, or cooktop can be cleaned with warm water, mild detergent, and a soft cloth on all cleanable parts, as recommended in the use and care manual. Never use scouring pads on these surfaces, except where recommended in the use and care manual. Also, never use abrasive cleaners that are not specifically recommended by the manufacturer.

Do not allow grease spillovers to accumulate after cooking on top of the range; it will become a fire hazard.

On glass surfaces, you can use a glass cleaner to clean any soil stains. Stubborn soil stains on glass surfaces can be removed with a paste of baking soda and water. Never use abrasive cleaners on glass surfaces.

Never clean the heating elements. When you turn on the elements, the soil will burn off.

On self-cleaning models, never use oven cleaners to clean the oven cavity. This will cause hazardous fumes when the oven is in the cleaning cycle. After the oven cavity has cooled, use only soap and water to clean small spills on the inside of the cavity.

REPAIR PROCEDURES

Each repair procedure is a complete inspection and repair process for a single range, oven, or cooktop component. It contains the information you need to test and replace components.

Infinite (burner) switch

The typical complaints associated with infinite (burner) switch failure are: 1. The surface element will not heat at all. 2. The surface element will stay on high in any position. 3. Intermittent switch operation.

1. *Verify the complaint* Verify the complaint by turning the infinite switch on. Is the surface element heating?
2. *Check for external factors* You must check for external factors not associated with the appliance. Is the appliance installed properly? Does the appliance have the correct voltage?
3. *Disconnect the electricity* Before working on the range or cooktop, disconnect the electricity to the appliance. This can be done by pulling the plug from the receptacle. Or disconnect the electricity at the fuse panel or at the circuit breaker panel. Turn off the electricity.
4. *Gain access to the infinite switch* Depending on which model you are repairing, you can access the component by removing the back panel (Fig. 12-6). On models with front-mounted controls (Fig. 12-7), the panel is

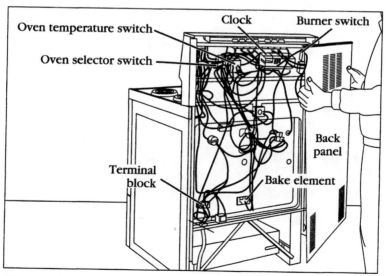

12-6 Removing the screws from the back panel to gain access to the components.

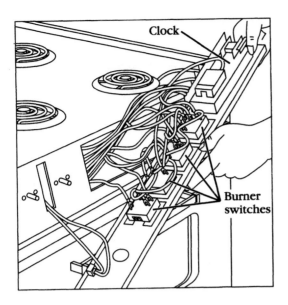

12-7
Removing the screws from both ends to remove the control panel and gain access to the components.

Clock

Burner switches

attached with screws on both ends. Remove the screws, and tilt the control panel. Be very careful not to let the wires disengage from the components. Some built-in models have a removable backsplash (Fig. 12-8); just lift the backsplash, and rest it on the cooktop. It would be a good idea to place something on the cooktop to protect it from getting damaged. Next, remove the screws from the backsplash, which hold the rear panel, to access the components. If you are repairing a wall oven or an eye-level range, the control panel can be removed (Fig. 12-9) by opening the door and removing the screws that secure the panel. The screws that secure the control panel to the unit might be underneath the front of the exhaust hood

12-8
Removing the screws from the back panel to gain access to the components.

Backsplash

End cap

12-9 Removing the screws that secure the control panel to the oven to gain access to the components.

or just below the control panel. Some control panels are hinged, just tilt the control panel toward you for servicing. To access the components on other models, both the rear panel and the front control panel (usually glass) will have to be removed. To pull out the glass, remove the screws that secure the trim piece that holds the glass in place.

5. *Test the infinite switch* To test the infinite switch for continuity between the switch contacts, remove all of the wires from the switch terminals (label them). Set the infinite switch in the "high" position. Using your ohmmeter, place the probes on the L1 and H1 (Fig. 12-10A) terminals; there should be continuity. Next, place the probes on the L2 and H2 (Fig. 12-10B) terminals, there should be continuity. Now, place the probes on the P and H1 (Fig. 12-10C) terminals, there should be continuity. The infinite switch is defective if there is no continuity between L1 and H1, or between L2 and H2. If there is no continuity between P and H1, the indicator light circuit is defective.

6. *Remove the infinite switch* To remove the infinite switch on most models, you must remove the two screws that secure the switch to the control panel. On some models, the infinite switch is secured with a nut to the control panel. Unscrew the nut to remove this type of infinite switch.

7. *Install a new infinite switch* To install the new infinite switch, just reverse the disassembly procedure, and reassemble. Be sure that you install the wires onto the correct terminals, according to the wiring diagram.

Surface (burner) element

The typical complaints associated with surface element (burner) failure are: 1. The surface element will not heat at all. 2. When the surface element is turned on, it trips

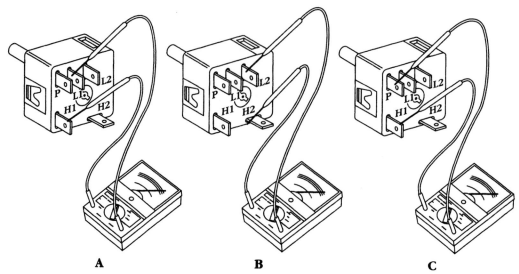

12-10 Testing an infinite switch for continuity.

the circuit breaker, or blows the fuses. 3. On dual surface elements, only part of the element heats up.

1. *Verify the complaint* Verify the complaint by turning on the infinite switch. Is the surface element heating?

2. *Check for external factors* You must check for external factors not associated with the appliance. Is the appliance installed properly, and does the appliance have the correct voltage?

3. *Disconnect the electricity* Before working on the range or cooktop, disconnect the electricity to the appliance. This can be done by pulling the plug from the receptacle. Or disconnect the electricity at the fuse panel or at the circuit breaker panel. Turn off the electricity.

4. *Remove the surface element* On most models, the surface element can be removed by simply pulling the element out of its receptacle (Fig. 12-11). Other models have the surface elements connected to the cooktop, and directly wired to their infinite switch wires (Fig. 12-12). To remove this type of element, remove the screw that secures the element to the cooktop. Then, remove the clips that hold the insulators to the element terminals (see inset). Next, unscrew the wires from the terminals; without bending those terminals.

5. *Test the surface element* Using the ohmmeter, set the range scale on R × 1. Place the probes on the element terminals (Fig. 12-13), the meter reading should show continuity. The actual readings will vary from manufacturer to manufacturer; they vary according to the size of the unit and the wattage used, but the readings should be generally between 19 and 115 ohms. To test for a grounded surface element, place one probe on the sheath, and the other probe on each element terminal, in turn (Fig. 12-14). If continuity exists at either terminal, the element is shorted, and it should be replaced.

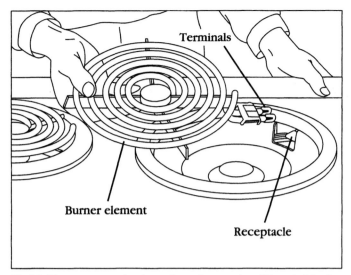

Terminals

Burner element

Receptacle

12-11 To remove a plug-in surface element, just lift and slide the element out of the receptacle.

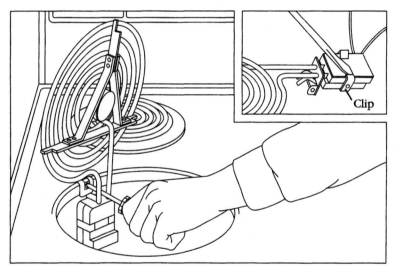

Clip

12-12 To remove a wired-in surface element, remove the screw that se-cures the element to the cooktop. Then, remove the insulator by prying off the clip.

6. *Install a new surface element* To install the new surface element, just reverse the disassembly procedure, and reassemble. Be sure that you install the wires on their correct terminals, according to the wiring diagram.

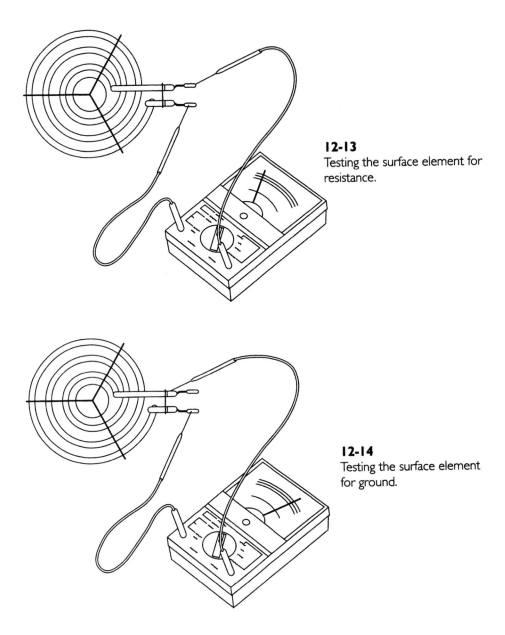

12-13
Testing the surface element for resistance.

12-14
Testing the surface element for ground.

Solid-disc elements

The typical complaints associated with solid-disc elements are the same as those for the traditional surface elements.

1. *Verify the complaint* Verify the complaint by turning the infinite switch on. Is the solid-disc element heating?
2. *Check for external factors* You must check for external factors not associated with the appliance. Is the appliance installed properly? Does the appliance have the correct voltage?

3. *Disconnect the electricity* Before working on the range or cooktop, disconnect the electricity from the appliance. This can be done by pulling the plug from the receptacle. Or disconnect the electricity at the fuse panel or to the circuit breaker panel. Turn off the electricity.

4. *Gain access to the solid-disc element* To access the element on this model, you will have to raise the cooktop. To do this, remove the screws from under the front edge of the cooktop (Fig. 12-15), then lift the cooktop and prop it up. On some models, the trim might have to be removed first.

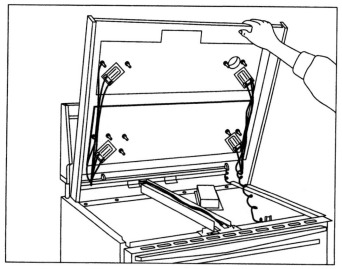

12-15 Removing the screws from under the front edge of the cooktop to gain access to the solid-disc elements.

5. *Test the solid-disc element* Remove the wires from the element. Using the ohmmeter, and set the range scale on R × 1. Place the probes on the element terminals; there should be continuity. The actual readings will vary from manufacturer to manufacturer; and with the size of the unit and the wattage used; but the readings should generally be between 19 and 115 ohms. To test for a grounded element, place one probe on the sheath, and attach the other probe to each element terminal. If continuity exists at either terminal, the element is shorted and it should be replaced.

6. *Remove the solid-disc element* To remove the solid-disc element, remove the wires from the element (Fig. 12-16). Remove any brackets that secure the element to the cooktop. The element is removed from the top of the cooktop (see inset).

7. *Install a new solid-disc element* To install the new solid-disc element, just reverse the disassembly procedure, and reassemble. Be sure that you install the wires on their correct terminals, according to the wiring diagram.

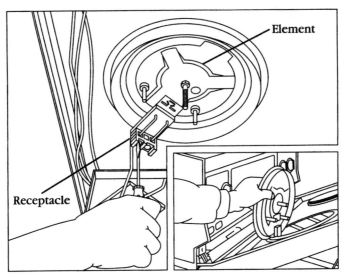

12-16 Removing the wires from the solid-disc element before you test for continuity. Remove the screws that secure the element, then remove the element from the top of the cooktop.

Radiant heating element

The typical complaints associated with the radiant elements are the same as those for surface elements.

1. *Verify the complaint* Verify the complaint by turning the infinite switch on. Is the radiant element heating?

2. *Check for external factors* You must check for external factors not associated with the appliance. Is the appliance installed properly, and does the appliance have the correct voltage?

3. *Disconnect the electricity* Before working on the range or cooktop, disconnect the electricity to the appliance. This can be done by pulling the plug from the receptacle. Or disconnect the electricity at the fuse panel or at the circuit breaker panel. Turn off the electricity.

4. *Gain access to the radiant heating element* To access the element, you will have to raise the cooktop (Fig. 12-15). On some models, you will have to remove the trim first. Remove the screws from under the front edge of the cooktop; then, lift the cooktop and disconnect the element's wiring harness plug. Remove the cooktop, and place it upside-down on a table. Don't forget to place a blanket under the cooktop to prevent the cooktop from being damage.

5. *Test the radiant heating element* In order to test the element, remove the element mounting bracket from the cooktop. This is held on with two screws on either end of the bracket. Then, carefully turn over the heating element assembly. This will give you access to the limiter and the element (Fig. 12-17). Be careful, the element is embedded in an insulated casting, which is used to prevent the cooktop glass from overheating. Remove the wires from the element. On the ohmmeter, set the range scale on R × 1.

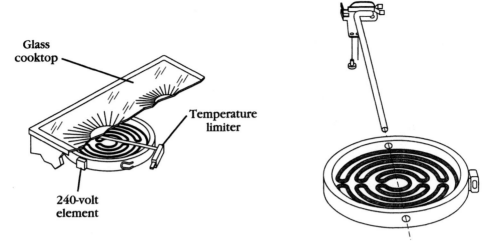

Glass cooktop

Temperature limiter

240-volt element

12-17 The radiant heating element must be flat against the glass.

Place the probes on the element terminals; there should be continuity. The actual readings will vary from manufacturer to manufacturer. They will vary according to the size of the unit and the wattage used, but the readings should generally be between 19 ohms and 115 ohms. If the element fails the test, replace it along with the limiter.

6. *Remove the radiant heating element* To remove the element, remove the screws that secure the element to the bracket. Lift the element out. The new element comes as a complete assembly.

7. *Install a new radiant heating element* To install the new radiant heating element, just reverse the disassembly procedure, and reassemble. The element must be flat against the glass and located under the heater panel (Fig. 12-17). Be sure that you install the wires on their correct terminals, according to the wiring diagram.

Bake element

The typical complaints associated with bake element failure are: 1. The bake element will not heat at all. 2. When the bake element is turned on, it trips the circuit breaker, or it blows the fuses.

1. *Verify the complaint* Verify the complaint by turning the selector switch to "bake" and setting the thermostat. Is the bake element heating? On some models, the clock must be set to "manual." Check the use and care manual for the model you are servicing.

2. *Check for external factors* You must check for external factors not associated with the appliance. Is the appliance installed properly? Does the appliance have the correct voltage?

3. *Disconnect the electricity* Before working on the range, disconnect the electricity to the range. This can be done by pulling the plug from the receptacle. Or disconnect the electricity at the fuse panel or at the circuit breaker panel. Turn off the electricity.

4. *Gain access to the bake element*   To access the bake element, open the oven door and remove the oven racks. Begin by removing the two screws that secure the element to the oven cavity (Fig. 12-18).

5. *Remove and test the bake element*   Slide the element forward and remove the wires from the bake element terminals (label them) either by removing the screws from the terminals, or by pulling the wires off of the bake element terminals (Fig. 12-19). Using the ohmmeter, set the range scale on R × 1. Place the probes on the element terminals (Fig. 12-20); there should be continuity. The actual readings will vary from manufacturer to manufacturer, according to the size of the unit and the wattage used; but the readings should generally be between 19 and 115 ohms. To test for a grounded bake element, place one probe on the sheath and the other probe on an element terminal (Fig. 12-21). If continuity exists, the element is shorted and it should be replaced. Test both terminals.

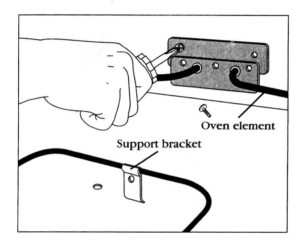

Oven element

Support bracket

12-18
Removing the two screws and sliding the element forward to get at the element terminals.

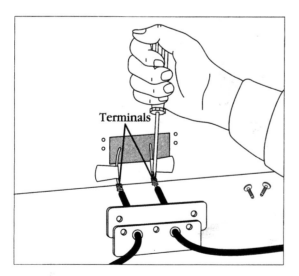

Terminals

12-19
Removing the screws and wires from the element before testing.

280   Electric ranges and ovens

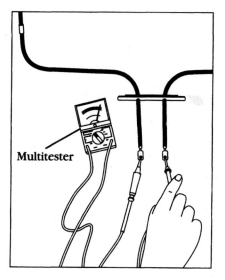

12-20
Testing the bake element for resistance.

Multitester

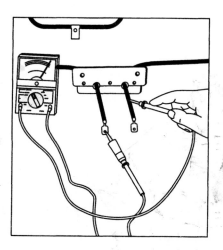

12-21
Testing the element for ground.

6. *Install a new bake element* To install the new bake element, just reverse the disassembly procedure, and reassemble. Be sure that you install the wires on their correct terminals, according to the wiring diagram.

Broil element

The typical complaints associated with broil element failure are: 1. The broil element will not heat at all. 2. When the broil element is turned on, it trips the circuit breaker or it blows the fuses. 3. On dual broil elements, only half of the element heats.

1. *Verify the complaint* Verify the complaint by turning the selector switch to "broil," and/or by setting the thermostat to "broil." Is the broil element heating? On some models, the clock must be set to "manual." Check the use and care manual for the model you are servicing.

2. *Check for external factors* You must check for external factors not associated with the appliance. Is the appliance installed properly? Does the appliance have the correct voltage?

3. *Disconnect the electricity* Before working on the range, disconnect the electricity. This can be done by pulling the plug from the receptacle. Or disconnect the electricity at the fuse panel or at the circuit breaker panel. Turn off the electricity.

4. *Gain access to the broil element* To access the broil element, open the oven door and remove the oven racks. The broil element is located at the top of the oven cavity. Begin by removing the two screws that secure the element to the oven cavity. Then, remove the holding brackets from the element, and slide the element forward.

5. *Remove and test the broil element* Remove the wires from the broil element terminals, either by removing the screws from the terminals, or by pulling the wires off the broil element terminals (label them). Using the ohmmeter, set the range scale on R × 1. Place the probes on the element terminals (Fig. 12-22); there should be continuity. The actual readings will vary from manufacturer to manufacturer, and according to the size of the unit and the wattage used. The readings should generally be between 19 and 115 ohms. To test for a grounded broil element, place one probe on the sheath and the other probe on an element terminal (Fig. 12-21). If continuity exists, the element is shorted and it should be replaced. Test both terminals.

   To test a dual broil element, remove the wires from their terminals. Using the ohmmeter, set the range scale on R × 1. Place the probes on terminals A and C in Figure 12-23; there should be continuity. Then, place the probes on terminals A and B (Figure 12-24); there should be continuity. Finally, place the probes on terminals B and C; there should be continuity. To test for a grounded element, place one probe on the sheath. With the other probe, touch terminal A, then B, then C (Fig. 12-25). If continuity exists, the element is shorted and it should be replaced.

6. *Install a new broil element* To install the new broil element, just reverse the disassembly procedure, and reassemble. Be sure that you install the wires on their correct terminals, according to the wiring diagram.

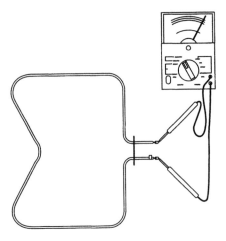

12-22
Testing the broil element for resistance.

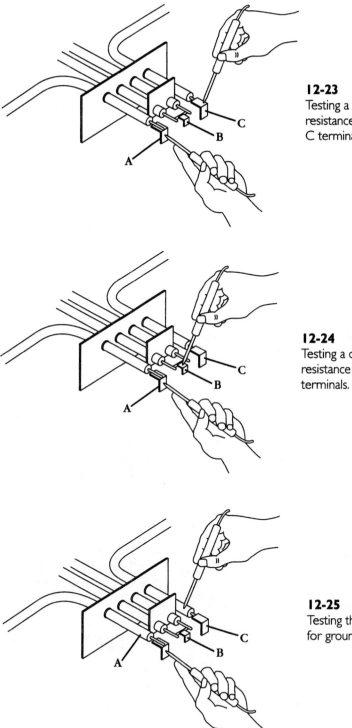

12-23
Testing a dual-broil element for resistance between the A and C terminals.

12-24
Testing a dual broil element for resistance between the A and B terminals.

12-25
Testing the dual broil element for ground.

Oven selector switch

The typical complaints associated with oven selector switch failure are: 1. Bake element will not heat. 2. Broil element will not heat. 3. Self-clean will not work. 4. Time bake will not work.

1. *Verify the complaint* Verify the complaint by turning the oven selector switch on to the desired setting. Set the thermostat to the desired temperature setting. Does that part of the range/oven work? Test all of the cycles on the selector switch.

2. *Check for external factors* You must check for external factors not associated with the appliance. Is the appliance installed properly? Does the appliance have the correct voltage?

3. *Disconnect the electricity* Before working on the range or oven, disconnect the electricity to the appliance. This can be done by pulling the plug from the receptacle. Or disconnect the electricity at the fuse panel or at the circuit breaker panel. Turn off the electricity.

4. *Gain access to the oven selector switch* Depending on which model you are repairing, you can access the component by removing the back panel (Fig. 12-6) on a freestanding range. On models with front-mounted controls (Fig. 12-7), the panel is attached with screws on both ends. Remove the screws, and tilt the control panel. Be very careful not to let the wires come off from their components. Some built-in models have a removable backsplash (Fig. 12-8). Lift the backsplash and rest it on the cooktop. It would be a good idea to place something on the cooktop first, to protect it from getting damaged. Next, remove the screws from the backsplash that hold the rear panel, to gain access to the components. If you are repairing a wall oven or an eye-level range, the control panel can be removed (Fig. 12-9) by opening the door and removing the screws that secure the panel. The screws that secure the control panel to the unit might be underneath the front of the exhaust hood or just below the control panel. Some control panels are hinged; just tilt the control panel towards you for servicing. On still other models, both the rear panel and the front control panel (usually glass) will have to be removed, to gain access to the components. To remove the glass, there are screws that secure the trim piece that holds the glass in place. Remove the screws and the trim ring. On some models, the trim might have to be removed first.

5. *Test the oven selector switch* To test the oven selector switch, for continuity between certain switch contacts, remove only those wires that are being tested (label them) (Fig. 12-26). To test each switch contact, refer to the wiring diagram for the proper switch contact terminals. Only remove one pair of wires at a time to test for continuity, as the oven selector switch is checked in each position.

6. *Remove the oven selector switch* To remove the oven selector switch on most models, you must remove the two screws that secure the switch to the control panel. On some models, the oven selector switch is secured to the control panel with a nut. Unscrew this nut to remove the oven selector switch (Fig. 12-27).

7. *Install a new oven selector switch* To install the new oven selector switch, just reverse the disassembly procedure, and reassemble. Be sure that you install the wires on the correct terminals, according to the wiring diagram.

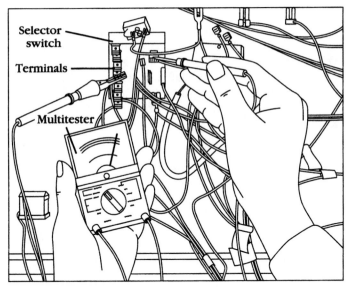

12-26 When testing the oven selector switch, only remove one wire from each pair of terminals being tested.

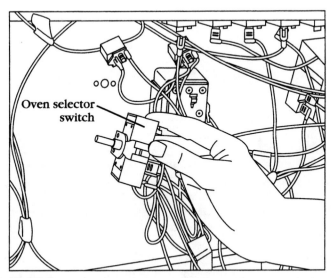

12-27 Removing the two screws that secure the oven selector switch to the control panel. At this time, do not remove the wires until the new switch is ready for installation. Then, transfer the wires to the correct terminals.

Oven thermostat

The typical complaints associated with oven thermostat failure are: 1. Bake element will not heat. 2. Broil element will not heat. 3. Timed-bake will not work. 4. Oven temperature not accurate.

1. *Verify the complaint* Verify the complaint by turning on the oven selector switch to the desired setting and by setting the oven thermostat to the temperature selected. Does that function of the range/oven work? On some models, the clock must be set to "manual." Check the use and care manual for the model you are servicing.

2. *Check for external factors* You must check for external factors not associated with the appliance. Is the appliance installed properly? Does the appliance have the correct voltage?

3. *Disconnect the electricity* Before working on the range or oven, disconnect the electricity to the appliance. This can be done by pulling the plug from the receptacle. Or disconnect the electricity at the fuse panel or at the circuit breaker panel. Turn off the electricity.

4. *Gain access to the oven thermostat* You can gain access to the component by removing the back panel (Fig. 12-6) on a freestanding range. On models with front-mounted controls (Fig. 12-8), the panel is attached with screws on both ends. Remove the screws, and tilt the control panel. Be very careful not to let the wires come off of their components. Some built-in models have a removable backsplash (Fig. 12-8); lift the backsplash and rest it on the cooktop. It would be a good idea to place something on the cooktop first, to protect it from getting damaged. Next, remove the screws from the backsplash, which holds the rear panel, to access the components. If you are repairing a wall oven or an eye-level range, the control panel can be removed (Fig. 12-9) by opening the door and removing the screws that secure the panel. The screws that secure the control panel to the unit might be underneath the front of the exhaust hood or just below the control panel. Some control panels are hinged, just tilt the control panel towards you for servicing. On still other models, both the rear panel and the front control panel (usually glass) will have to be removed, to gain access to the components. To remove the glass, there are screws that secure the trim that holds the glass in place. Remove the screws and the trim piece. On some models, the trim might have to be removed first.

5. *Test the oven thermostat switch contacts* To test the oven thermostat switch for continuity between certain switch contacts, remove only those wires that are being tested from their terminals (Fig. 12-28). To test each switch contact, refer to the wiring diagram for the proper switch contact terminals. Only remove one pair of wires at a time (label them) to test for continuity.

6. *Calibrate the oven thermostat* Before making any adjustments to the thermostat, test the oven temperature. With an oven temperature tester, place the thermocouple tip in the center of the oven cavity. Be sure that the thermocouple tip does not touch any metal. Close the oven door; set the oven to bake; and adjust the thermostat setting to the 350-degree mark. Let the oven cycle for 20 to 30 minutes. Then, record the minimum and

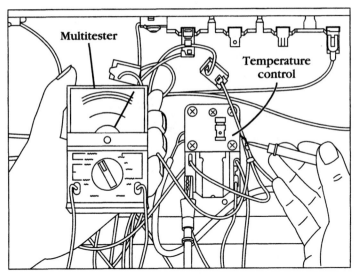

12-28 Testing the oven thermostat switch contacts for continuity.

maximum temperatures of three cycles. Next, add these temperatures, and divide by 6. This will give you the average temperature of the oven.

$$\frac{370 + 335 + 350 + 340 + 360 + 335}{6} = 348.3 \text{ degrees F.}$$

The average temperature calculated should be within ±25 degrees of the temperature setting selected (rotary dial type). If not, try calibrating the thermostat. To calibrate the thermostat, pull the oven thermostat knob off. Turn it over (Fig. 12-29). On the back of the knob is the calibration ring. Loosen the two screws, and move the pointer in the direction needed. Each line that the pointer is moved indicates 10 degrees of change in the calibration. Tighten the screws and place the knob back on the thermostat stem. Retest the oven temperature. On models that do not have the calibration ring on the back of the dial, replace the thermostat if the temperature is more than 25 degrees out of calibration.

7. *Remove the oven thermostat* To remove the oven thermostat on most models, you must disconnect the thermostat capillary tube from its supports (Fig. 12-30) and push it through the back wall of the oven cavity. Be careful. Do not break the capillary tube wire, because the contents inside are flammable. Next, remove the two screws that secure the thermostat to the control panel. Remove the thermostat (Fig. 12-31). Leave the wires on the thermostat for now.

8. *Install a new oven thermostat* To install the new oven thermostat, just reverse the disassembly procedure, and reassemble. Transfer the wires, one at a time, from the old thermostat to the new one. Be sure that you install the wires on their correct terminals, according to the wiring diagram. Reassemble the control panel in the reverse order of disassembly and test.

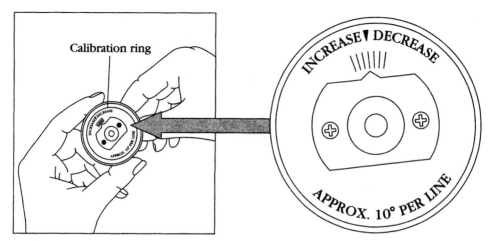

12-29 Turn the oven thermostat knob over. On the back of the knob are the words *increase* and *decrease*. By moving the pointer, you can increase or decrease the temperature setting on the thermostat. Each line indicates a 10° increment.

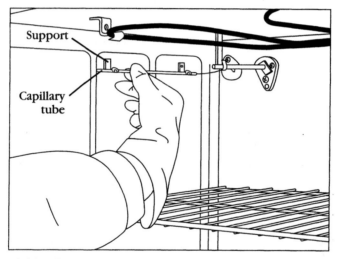

12-30 Lift the thermostat capillary tube off the supports. Be careful so that you don't break the tube when you slide it through the back of the oven cavity.

Oven cycling relay

The typical complaints associated with oven cycling relay failure are: 1. Bake element will not heat. 2. Broil element will not heat. 3. Bake element stays on all the time. 4. Oven temperature not accurate.

    1. Verify the complaint Verify the complaint by turning the oven selector switch to the desired setting, and by setting the oven thermostat to the temperature selected. Does that function of the oven work? Test all of the cycles.

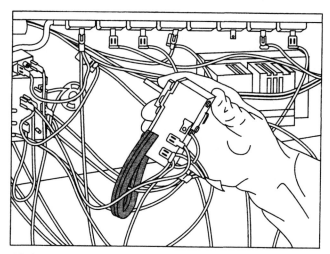

12-31 Removing the oven thermostat from the range. Leave the wires on the thermostat until the new thermostat is ready for installation. Then, transfer the wires to the correct terminals.

2. *Check for external factors* You must check for external factors not associated with the appliance. Is the appliance installed properly? Does the appliance have the correct voltage?

3. *Disconnect the electricity* Before working on the range or oven, disconnect the electricity to the appliance. This can be done by pulling the plug from the receptacle. Or disconnect the electricity at the fuse panel or at the circuit breaker panel. Turn off the electricity.

4. *Gain access to the oven cycling relay* To access the oven cycling relay (Fig. 12-32), remove the back panel (Fig. 12-6). On some models, this relay is located under the oven cavity, where the utility drawer is usually found.

5. *Test the oven cycling relay* To test the oven cycling relay, remove all of the wires from the relay (label them). Using the ohmmeter, set the range scale on R × 1. Place the probes on the L and H1 terminals; there should not be continuity. Next, place the probes on the L and H2 terminals; there should not be continuity. Next, place the probes on the H1 and H2 terminals. There should not be continuity. Now, place the probes on the S and R terminals. There should be continuity. If your meter indicates no reading at all between the terminals R and S, this indicates that the heater wire is defective inside the relay, and that the relay should be replaced.

6. *Remove the oven cycling relay* To remove the oven cycling relay, remove the two screws that secure the relay to the cabinet.

7. *Install a new oven cycling relay* To install the new oven cycling relay, just reverse the order of step 6. Transfer the wires from the old relay to the new relay. Be sure that you install the wires on their correct terminals, according to the wiring diagram. Reassemble the remainder of the range/oven in the reverse order of its disassembly and test.

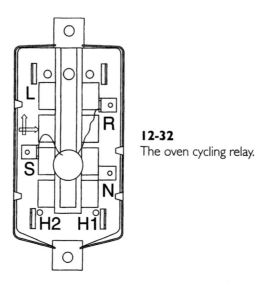

12-32
The oven cycling relay.

Range/oven time clock

The typical complaints associated with range/oven time clock failure are: 1. Timed-bake cycle will not operate. 2. Unable to use the self clean cycle. 3. Clock loses time. 4. Clock is not functioning. 5. The minute reminder is not functioning.

1. *Verify the complaint* Verify the complaint by turning the clock controls to the correct time. Does the clock time advance? Some models have the minute reminder located on the same clock, set this timer to test it. Does it advance?

2. *Check for external factors* You must check for external factors not associated with the appliance. Is the fuse blown? Does the appliance have the correct voltage?

3. *Disconnect the electricity* Before working on the range or oven, disconnect the electricity to the appliance. This can be done by pulling the plug from the receptacle. Or disconnect the electricity at the fuse panel or at the circuit breaker panel. Turn off the electricity.

4. *Gain access to the range/oven clock* You can access the clock by removing the back panel (Fig. 12-6) on a freestanding range. On models with front-mounted controls (Fig. 12-7), the panel is attached with screws on both ends. Remove the screws, and tilt the control panel. Be very careful not to let the wires come off their components. Some built-in models have a removable backsplash (Fig. 12-8); just lift the backsplash and rest it on the cooktop. It would be a good idea to place something on the cooktop first, to protect it from getting damaged. Next, remove the screws from the backsplash, that hold the rear panel, to gain access to the clock. If you are repairing a wall oven, or an eye-level range, the control panel can be removed (Fig. 12-9) by opening the door and removing the screws that secure the panel. The screws that secure the control panel to the unit might be underneath the front of the exhaust hood or just below the control panel. Some control panels are hinged; just tilt the control panel towards

you for servicing. On still other models, to gain access to the clock, both the rear panel and the front control panel (usually glass) will have to be removed. There are screws that secure the trim piece that holds the glass in place. Remove the screws. Figure 12-33 illustrates the different types of clock faces available on ranges/ovens.

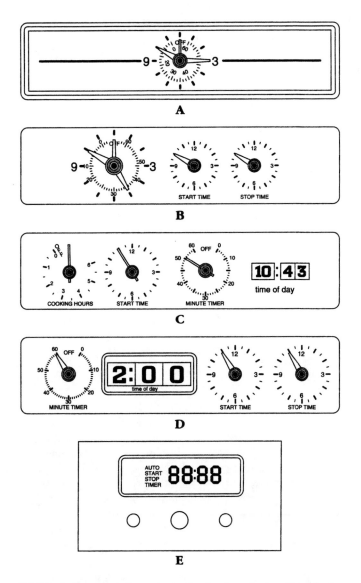

12-33 A. Standard time clock. B. Analog clock with minute reminder, time bake, and self clean. C and D. Digital clock with time bake, self clean, and minute reminder. E. Electronic clock, with or without additional features.

5. *Test the range/oven clock* On some models, if the clock does not run, check for a fuse in the circuit. Locate the clock motor wire leads, and isolate them from the circuit (label them). Using the ohmmeter, set the range scale on R × 1. Place the probes on the clock motor leads; there should be continuity. If not, replace the clock. On some models, the clock is used to control the timed-bake and the self-clean cycles (Figs. 12-33B, C, and D). To check the switch mechanism of the clock, remove the wires from the switch terminals (label them). Using the ohmmeter, set the range scale on R × 1. Place the probes on the terminals. Look at the wiring diagram for the correct terminals to test. Some models have one switch, but other models have two sets of switches. Test for the continuity of the switch contacts when you push in and turn the start and stop knobs on the clock, and when the knobs pop out.
6. *Remove the range/oven time clock* First, remove the clock knobs from the clock stems. To remove the range/oven time clock in this model (Fig. 12-34), use a screwdriver and depress the clips that hold the clock to the control panel. On other models, the clock is secured to the control panel by screws or nuts. Pull the clock toward the front of the appliance (Fig. 12-35).
7. *Install a new range/oven clock* Transfer the wires from the old clock to the new clock. Be sure that you install the wires on their correct terminals, according to the wiring diagram. To install the new range/oven clock, just reverse the disassembly procedure, and reassemble.

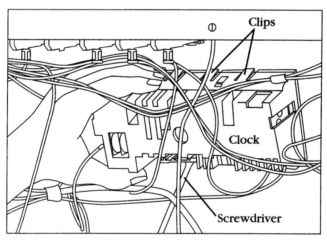

12-34 Clocks are secured to the control panel with clips, screws, or nuts.

DIAGNOSTIC CHARTS AND WIRING DIAGRAMS

The following diagnostic charts will help you to pinpoint the likely causes of the various problems (Figs. 12-36, 12-37, 12-38, and 12-39).

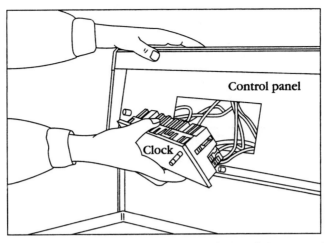

12-35 Removing the clock from the front of the control panel.

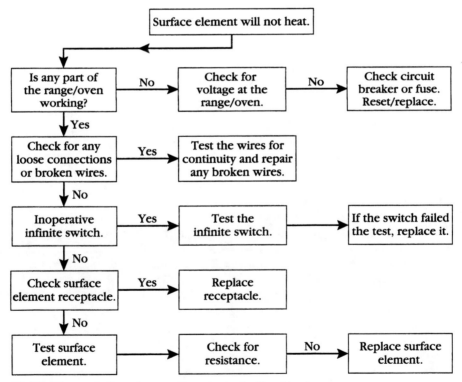

12-36 Diagnostic flow chart: surface element will not heat.

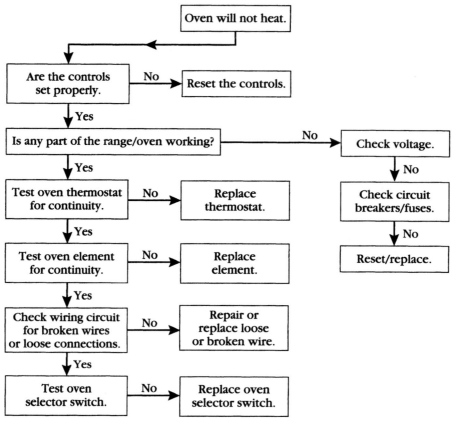

12-37 Diagnostic flow chart: oven will not heat.

The wiring diagrams in this chapter are examples only. You must refer to the wiring diagram on the range, oven, or cooktop that you are servicing. Figures 12-40 and 12-41, are for an identical appliance. Figure 12-40 depicts an actual wiring schematic. Figure 12-41 depicts a cross section of a ladder wiring diagram. This type of diagram is easier to read and understand.

Table 12-1, illustrates a self-clean cycle chart, indicating a light soil cycle. On some models, the cleaning cycle will take two hours to complete. The heavy soil cycle will take two to three hours to complete. The average stabilized temperature in a self-cleaning cycle should be between 840 and 920 degrees Fahrenheit. The door can only be opened when the oven temperature has dropped below 520 degrees Fahrenheit.

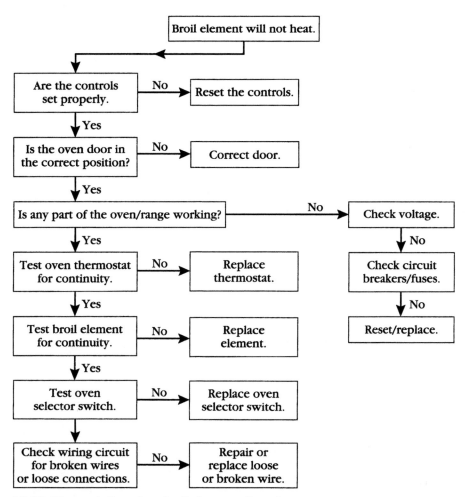

12-38 Diagnostic flow chart: broil element will not heat.

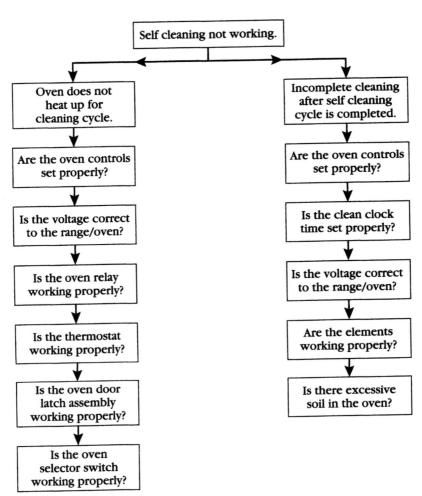

12-39 Diagnostic flow chart: self-cleaning not working.

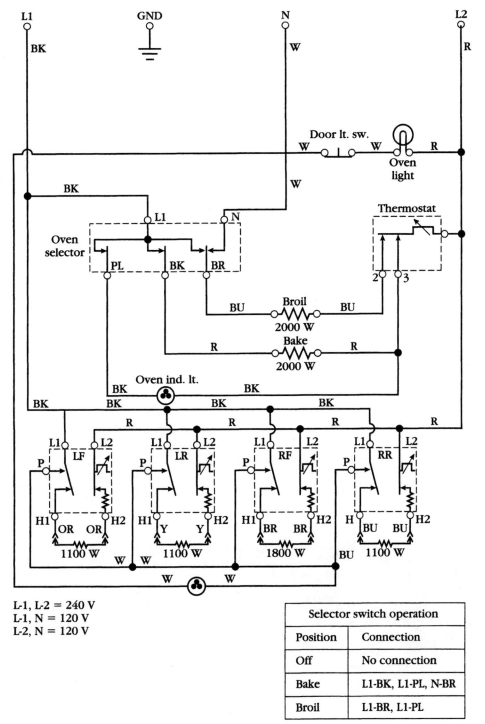

L-1, L-2 = 240 V
L-1, N = 120 V
L-2, N = 120 V

Selector switch operation	
Position	Connection
Off	No connection
Bake	L1-BK, L1-PL, N-BR
Broil	L1-BR, L1-PL

12-40 Wiring schematic of a typical electric range.

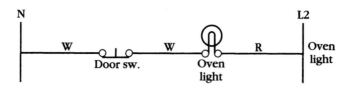

Example only: refer to the diagram on the unit when servicing
L-1, L-2 = 240 V
N, L-1 = 120 V
N, L-2 = 120 V

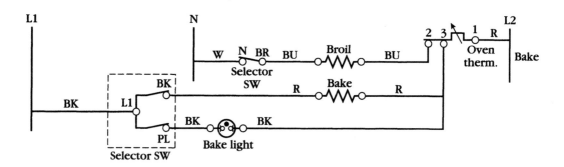

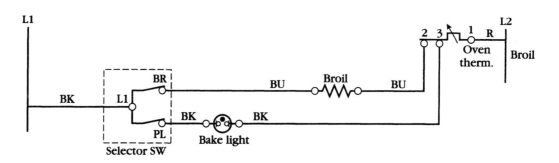

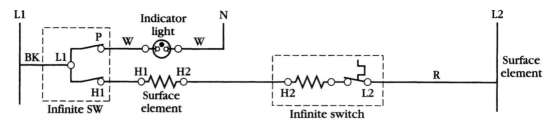

12-41 The oven light, bake, broil, and surface element circuits.

TABLE 12-1

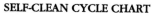

SELF-CLEAN CYCLE CHART

Self-clean oven—Light soil cycle

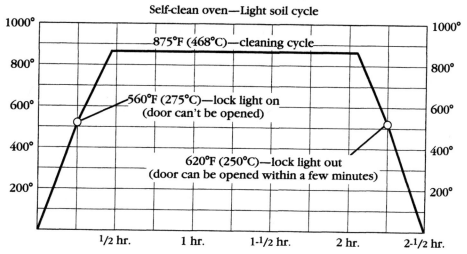

13
Refrigerators and freezers

One of the most important applications of refrigeration (which was invented in the early 1900s) was for the preservation of food. When different types of food[1] are kept at room temperature, some of them will spoil rapidly. When foods are kept cold, they will last longer. Refrigerator/freezers prevent food spoilage by keeping the food cold.

The refrigerator/freezer consists of three parts. They are:
1. The cabinet.
2. The sealed system, which consists of the evaporator coil, the condenser coil, the compressor, and the connecting tubing.
3. The electrical circuitry.

THE REFRIGERATION CYCLE

The sealed system (Fig. 13-1) in a refrigerator or freezer consists of a compressor, a condenser coil, an evaporator coil, a capillary tube, and a heat exchanger and its connecting tubing. This is the heart of the refrigerator or freezer that keeps the food cold inside of the cabinet.

Starting at the compressor, refrigerant gas is pumped out of the compressor, through the discharge tubing, and into the condenser coil. When the gas is in the condenser coil, the temperature and pressure of the refrigerant gas greatly increases because of the capillary tube at the discharge end of the condenser coil. From the surface of the condenser coil, the heat spreads out into the room via air moving over the condenser coil. The condenser coil cools the hot refrigerant gas. As the refrigerant gas gives up the heat it obtained from inside the refrigerator or freezer cabinet, the refrigerant gas changes into a liquid. This liquid then leaves the condenser coil and enters the capillary tube.

This capillary tube is carefully made, as to its length and inside diameter, to meter the exact amount of liquid flow through the sealed system (as designed by the

---

[1] Dairy products, meats, seafood, fruits, and vegetables will all spoil rapidly, if not kept cold or frozen. The colder temperatures of the refrigeration compartment, between 35 and 40 degrees Fahrenheit, in a refrigerator will slow the spoiling of foods. Most foods will last from 3 to 7 days at that temperature. If the foods were frozen, and packaged properly, they could last for several weeks in a domestic refrigerator with a temperature of 0 degrees Fahrenheit to −10 degrees Fahrenheit.

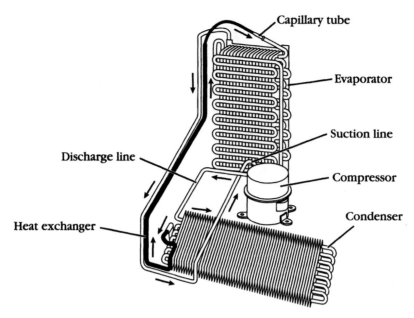

13-1 The sealed-system component location for a side-by-side refrigerator.

manufacturer, for a particular size and model). As the liquid refrigerant leaves the capillary tube and enters the larger tubing of the evaporator coil, the sudden increase of tubing size causes a low-pressure area. It is here that the liquid refrigerant changes from a liquid to a mixture of liquid and gas.

As this mixture of liquid and gas passes through the evaporator coil, the refrigerant absorbs heat from the warmer items (food) within the refrigerator or freezer cabinet, slowly changing any liquid back to all gas. As the refrigerant gas leaves the evaporator coil, it returns to the compressor through the suction line.

This entire procedure is called a *cycle*. Depending on where the cold control is set, the thermostat can show how cold it is inside the cabinet, and then control the actuation of the cooling cycle. It will determine whether to turn the system on or off, to maintain the temperature within the cabinet.

Inside the cabinet, the cold air is circulated by convection and/or by means of an electrical fan. In Fig. 13-2, the arrows are showing the air flow patterns in this type of side-by-side refrigerator. Figure 13-3 shows the air patterns in this type of two-door refrigerator with a top freezer.

Storage requirements for perishable products

Table 13-1 represent the recommended storage temperatures, relative humidity, and the approximate storage life for perishable products. These values are used in designing commercial refrigeration systems, which house large quantities of perishable products. Large warehouses are usually equipped to store foods at those temperatures best adapted to prolong the safe storage period for each type of food. In the domestic refrigerator, most foods are kept at 35 to 45 degrees Fahrenheit, and the humidity is kept around 50%. The freezer temperature is between zero and 10 degrees Fahrenheit. It

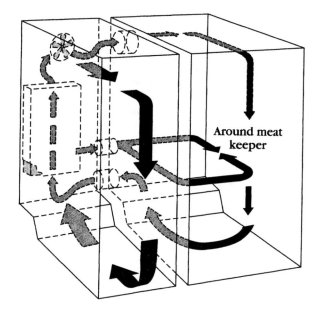

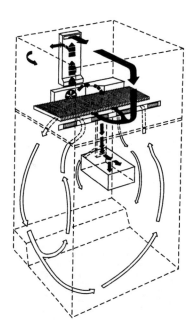

Around meat
keeper

13-2
The air-flow pattern in a side-by-side refrigerator.

13-3
The air-flow pattern in a two-door refrigerator with top freezer.

will be difficult to maintain these temperatures and humidities for each individual product. Refrigerator manufacturers have designed separate compartments, within the refrigerated cabinet, to maintain a variable temperature and humidity, as selected by the consumer. The storage life of various products will vary in a domestic refrigerator/freezer. This period will be influenced by many factors, such as the storage temperature, the type of container, the condition of the food, and the kind of food.

Table 13-1. Storage requirements for perishable products

Product	Storage Temp. F.	Relative Humidity %	Approximate Storage Life
Apples	30–40	90	3–8 months
Apricots	31–32	90	1–2 weeks
Artichokes	31–32	95	2 weeks
Asparagus	32–36	95	2–3 weeks
Avocados	45–55	85–90	2–4 weeks
Bananas	55–65	85–95	- - -
Beans (green or snap)	40–45	90–95	7–10 days
Beans, lima	32–40	90	1 week
Blackberries	31–32	95	3 days
Blueberries	31–32	90–95	2 weeks
Broccoli	32	95	10–14 days
Cabbage	32	95–100	3–4 months
Carrots	32	98–100	5–9 months
Cauliflower	32	95	2–4 weeks
Celery	32	95	1–2 months
Cherries, sour	31–32	90–95	3–7 days
Cherries, sweet	30–31	90–95	2–3 weeks
Collards	32	95	10–14 days
Corn, sweet (fresh)	32	95	4–8 days
Cranberries	36–40	90–95	2–4 months
Cucumbers	50–55	90–95	10–14 days
Dairy products			
Cheddar cheese	40	65–70	6 months
Processed cheese	40	65–70	12 months
Butter	40	75–85	1 month
Cream	35–40	- - -	2–3 weeks
Ice cream	−20 to −15	- - -	3–12 months
Milk, fluid whole			
Pasteurized, grade A	32–34	- - -	2–4 months
Condensed, sweetened	40	- - -	15 months
Evaporated	40	- - -	24 months
Dates (dried)	0 or 32	75 or less	6–12 months
Dried fruits	32	50–60	9–12 months
Eggplant	45–50	90–95	7–10 days
Eggs, shell	29–31	80–85	5–6 months
Figs, dried	32–40	50–60	9–12 months
Figs, fresh	31–32	85–90	7–10 days
Fish, fresh	30–35	90–95	5–15 days
Haddock, cod	30–35	90–95	15 days
Salmon	30–35	90–95	15 days
Smoked	40–50	50–60	6–8 months

Product	Storage Temp. F.	Relative Humidity %	Approximate Storage Life
Shellfish, fresh	30–33	86–95	3–7 days
Tuna	30–35	90–95	15 days
Grapefruit	50–60	85–90	4–6 weeks
Grapes, American type	31–32	85–90	2–8 weeks
Grapes, European type	30–31	90–95	3–6 months
Greens, leafy	32	95	10–14 days
Guavas	45–50	90	2–3 weeks
Honey	38–50	50–60	1 year, plus
Horseradish	30–32	95–100	10–12 months
Lemons	32 or 50–58	85–90	1–6 months
Lettuce, head	32–34	95–100	2–3 weeks
Limes	48–50	85–90	6–8 weeks
Maple sugar	75–80	60–65	1 year, plus
Mangoes	55	85–90	2–3 weeks
Meat			
Bacon, cured (farm style)	60-65	85	4–6 months
Game, fresh	32	80–85	1–6 weeks
Beef, fresh	32–34	88–92	1–6 weeks
Hams & shoulders, fresh	32–34	85–90	7–12 days
Cured	60–65	50–60	0–3 years
Lamb, fresh	32–34	85–90	5–12 days
Livers, frozen	–10–0	90–95	3–4 months
Pork, fresh	32–34	85–90	3–7 days
Smoked sausage	40–45	85–90	6 months
Fresh	32	85–90	1–2 weeks
Veal, fresh	32–34	90–95	5–10 days
Melons, Cantaloupe	36–40	90–95	5–15 days
Honeydew & Honey Ball	45–50	90–95	3–4 weeks
Watermelons	40–50	80–90	2–3 weeks
Mushrooms	32	90	3–4 days
Milk	34–40	- - -	7 days
Nectarines	31–32	90	2–4 weeks
Nuts (dries)	32–50	65–75	8–12 months
Okra	45–50	90–95	7–10 days
Olives, fresh	45–50	85–90	4–6 weeks
Onions (dry) & onion sets	32	65–70	1–8 months
Oranges	32–48	85–90	3–12 weeks
Orange juice, chilled	30–35	- - -	3–6 weeks
Papayas	45	85–90	1–3 weeks
Parsley	32	95	1–2 months
Parsnips	32	98–100	4–6 months
Peaches	31–32	90	2–4 weeks

Table 13-1. Continued

Product	Storage Temp. F.	Relative Humidity %	Approximate Storage Life
Pears	29–31	90–95	2–7 months
Peas, green	32	95	1–3 weeks
Peppers, sweet	45–50	90–95	2–3 weeks
Pineapples, ripe	45	85–90	2–4 weeks
Plums, including fresh prunes	31–32	90–95	2–4 weeks
Popcorn, unpopped	32–40	85	4–6 months
Potatoes, early crop	50–55	90	0–2 months
Potatoes, late crop	38–50	90	5–8 months
Poultry			
Fresh chicken	32	85–90	1 week
Fresh goose	32	85–90	1 week
Fresh turkey	32	85–90	1 week
Pumpkins	50–55	70–75	2–3 months
Radishes—spring, prepacked	32	95	3–4 weeks
Raisins (dried)	40	60–70	9–12 months
Rabbits, fresh	32–34	90–95	1–5 days
Raspberries, black	31–32	90–95	2–3 days
Raspberries, red	31–32	90–95	2–3 days
Rhubarb	32	95	2–4 weeks
Spinach	32	95	10–14 days
Squash, summer	32–50	85–95	5–14 days
Squash, winter	50–55	70–75	4–6 months
Strawberries, fresh	31–32	90–95	5–7 days
Sugar, maple	75–80	60–65	1 year, plus
Sweet potatoes	55–60	85–90	4–7 months
Syrup, maple	31	60–70	1 year, plus
Tangerines	32–38	85–90	2–4 weeks
Tomatoes, mature green	55–70	85–90	1–3 weeks
Tomatoes, firm ripe	45–50	85–90	4–7 days
Turnips, roots	32	95	4–5 months
Vegetables (mixed)	32–40	90–95	1–4 weeks
Yams	60	85–90	3–6 months

SAFETY FIRST

Any person who cannot use basic tools, or follow written instructions, should *not* attempt to install, maintain, or repair any refrigerators or freezers. Any improper installation, preventive maintenance, or repairs could create a risk of personal injury or property damage.

If you do not fully understand the installation, preventive maintenance, or repair procedures in this chapter, or if you doubt your ability to complete the task on your refrigerator or freezer, then please call your service manager.

Before continuing, take a moment to refresh your memory of the safety procedures in chapter 2.

This chapter covers the electrical components and how to diagnose the sealed system. The actual repair or replacement of any sealed system component is not included in this chapter. It is recommended that you acquire refrigerant certification (or call an authorized service company) to repair or replace any sealed system component. The refrigerant in the sealed system must be recovered properly.

REFRIGERATOR/FREEZERS IN GENERAL

Much of the troubleshooting information in this chapter covers refrigerators and freezers in general, rather than specific models, in order to present a broad overview of service techniques. The pictures and illustrations that are used in this chapter are for demonstration purposes to clarify the description of how to service these appliances. They in no way reflect on a particular brand's reliability.

Location and installation of refrigerator/freezer

Thoroughly read the installation instructions that are provided with every new refrigerator/freezer. These instructions will provide you with the information needed to properly install the refrigerator or freezer. Listed are some general principles that should be followed when performing the installation.

1. The refrigerator/freezer must be installed on a solid floor capable of supporting the product up to 1000 pounds.
2. For proper air circulation around the refrigerator/freezer, some models require a one-inch clearance at the rear and top of the cabinet, and adequate clearance near the front grille at the bottom of the refrigerator/freezer.
3. Do not leave the refrigerator/freezer on its side longer than necessary to remove the shipping base.
4. When removing or reversing the doors on a refrigerator, always reinstall them according to the installation instructions, and remember to realign the doors properly.
5. Level the refrigerator/freezer cabinet so that the doors will close properly.

What's that "different" sound in your kitchen?

If you have bought or serviced a new refrigerator within the past few years, you've probably noticed that it sounds "different." Here's why.

New refrigerators use only half as much electricity as the older models. In fact, a new 20.6-cubic foot refrigerator with top freezer uses no more electricity than a 75-watt light bulb. Most new refrigerators are also larger, and they have such added conveniences as automatic defrost systems, icemakers, and perhaps even a "built-in" look.

These new features result in different sounds, such as:

• High-pitched whine; from the more energy efficient compressors that have smaller, higher speed motors.

- Soft hum; from the evaporator fan in the freezer and/or from the condenser fan under the refrigerator.
- Clicks; from the automatic defrost timer, switching on and off; the thermostat, turning the refrigerator on and off; or the water valve, refilling the icemaker.
- Boiling and/or gurgling or trickling water; when the refrigerator stops running, the refrigerant continues to circulate within the system; or, the defrost water runs into the drain pan.
- Running water and "thuds;" as the ice cube tray fills, and as the ice cubes drop into the storage bin.

To help deafen these new sounds:
- Be sure that the refrigerator is level, and that the defrost water collection pan is in position (usually reachable behind the bottom front "toe plate").
- Put a piece of carpet or a sound-absorbing ceiling tile on the wall behind the refrigerator.
- Allow enough space between the back of the refrigerator and the wall, unless it is designed as a "built-in." Check your use and care book for the needed space.

To reduce the compressor "run time":
- Vacuum the condenser coils at least twice a year, more often if you have pets.
- Keep your freezer at least three-fourths full. Use partially filled water jugs to fill any large empty space.

Many of the consumers filing "refrigerator sound" complaints with MACAP represent one or two-family households; or they have recently moved to smaller retirement homes; or they have remodeled, with the kitchen open to a family living area. Sounds are most noticeable in quieter surroundings. Consumers with hearing aids are especially sensitive.

Some consumers report to MACAP that their refrigerators are "louder" than an identical model in a friend's or a relative's house. This might be because of the number of people in the house, as well as different furnishings and room arrangements. Carpeting, drapery, upholstered furniture, and wall coverings can help to muffle refrigerator sounds.[2]

Are refrigerator/freezers snowbirds?

Putting that extra refrigerator or freezer in any area in which the temperature falls below 60 degrees Fahrenheit might be a problem in colder areas when the winter months are approaching.

Combination refrigerator/freezers, and freezers with automatic defrost systems, are sensitive to the ambient air temperature surrounding them. As this ambient temperature rises, the compressor runs more to maintain the storage temperature in the fresh food and frozen food compartments, thus wasting energy. As the ambient temperature falls, compressor operation decreases.

When temperatures fall below 60 degrees Fahrenheit, the compressor will not operate long enough to maintain low storage temperatures in the freezer compartment. This is because the fresh food compartment contains the primary sensor, and

---

[2] Reprinted from MACAP Consumer Bulletin, issue no. 15, December 1992

it is satisfied quickly at a low ambient temperature. The lower the temperature goes, the worse that this condition becomes. At about 38 to 42 degrees Fahrenheit, the compressor will not run at all. The freezer compartment temperature will increase to the ambient room temperature, and the frozen food in the compartment will defrost and spoil.

Combination refrigerator/freezers and automatic defrost freezers should not be operated in unheated places like garages or porches, where room temperatures are likely to drop below 60 degrees Fahrenheit, unless they are specifically designed for operation in low temperatures. Check the manufacturer's use and care booklet for the lowest safe operating temperature.

At any time that the temperature will be less than 60 degrees Fahrenheit, it is best to empty the freezer compartment of the refrigerator/freezer, in order to prevent defrosting and the possible spoilage of the frozen foods. You might want to consider removing all of the food from the unit; turning it off; and propping the door open, if you will be absent for an extended period of time. The door must stay open to prevent mold and odor. *Don't* do this if your refrigerator has a latch-type handle (pre-1958 model) because of the potential for child entrapment.

Manual defrost freezers can usually be operated in an unheated garage or porch, without affecting the unit or the frozen food. However, check the use and care manual to determine if your unit needs special care[3].

"Freezer" in single-door refrigerator has limited function

"I just bought a single-door refrigerator and my ice cream won't harden. It's like cold soup," a consumer recently complained to MACAP. "If I turn the temperature control to a colder setting, the food in the refrigerator section freezes, but the ice cream remains soup. Something's wrong with the freezer."

Approximately 6% of all refrigerators sold each year are "single-door" models. That is, the model has only one outside door. Inside (usually across the top or to one side) is a small freezer compartment with its own door. Most combination "refrigerator/freezers" have two or more doors on the outside, providing independent access to separate freezer or refrigerated sections.

In the course of investigating this consumer's complaint, MACAP found that consumers have reason to be confused about the capabilities of "single-door" refrigerators. Many manufacturers refer to the separate frozen food compartment as a "freezer section," as a "freezer compartment," or as a "freezer" in their specification literature. MACAP has found that a majority of consumers expect to be able to keep such hard-to-freeze items as ice cream and orange juice in this "freezer" compartment. These items have a high sugar content, and they freeze at lower temperatures than water.

The Association of Home Appliance Manufacturers (AHAM), in its nationally accepted standard, calls such a unit a "basic refrigerator," and specifies that it is "intended for short-term storage of foods at temperatures below 32 degrees Fahrenheit and normally above 8 degrees Fahrenheit." However, most newer models have temperatures at (or near) freezing level, which is not adequate to freeze foods. Distilled water freezes at 32 degrees Fahrenheit, but all frozen foods must be stored at a tem-

---

[3] Reprinted from MACAP Consumer Bulletin, issue no. 12, November 1989

perature lower than that to freeze. Vegetables begin to freeze at 29 to 31 degrees Fahrenheit, meats at 25 to 29 degrees Fahrenheit, and orange juice concentrate at about 8 degrees Fahrenheit. Ice cream begins to stiffen at 27 degrees Fahrenheit, but is considered at an ideal hardness for scooping at 8 degrees Fahrenheit.

If reference is made to the compartment as being a "food freezer" or as a "frozen food storage compartment" (as in double-door units), the consumer can expect that it will store already frozen foods for several days, or even months, without deterioration. However, a "freezer section" or "freezer compartment," as found in single-door refrigerators, might only freeze ice cubes.

MACAP recommends that consumers determine their food freezing needs and carefully read available literature before making a purchasing decision.[4]

REFRIGERATOR/FREEZER MAINTENANCE

The inside of the cabinet should be cleaned at least once a month, to help prevent odors from building up. Of course, any spills that might happen should be wiped up immediately. Wash all removable parts by hand with warm water and a mild detergent; then rinse and dry the parts. The inside walls of the cabinet, the door liners, and the gaskets should also be washed, using warm water and a mild detergent, rinsed and dried. Never use cleaning waxes, concentrated detergents, bleaches, or cleansers containing petroleum products on plastic parts. On the outside of the cabinet, use a sponge with warm water and a mild detergent to clean dust and dirt. Then, rinse off and dry thoroughly.

At least two times a year, the outside cabinet should be waxed with an appliance wax, or with a good auto paste wax. Waxing painted metal surfaces provides rust protection. The defrost pan, which is located behind the toe plate or behind the cabinet, should be cleaned out once a month. The condenser coil should also be cleaned of dust and lint at least once a month also. The floor should be free of dirt and debris when the cabinet is rolled out away from the wall. After the cabinet is rolled back into place, you must check to be sure that the cabinet is level.

STEP-BY-STEP TROUBLESHOOTING BY SYMPTOM DIAGNOSIS

When servicing an appliance, don't overlook the simple things that might be causing the problem. Step-by-step troubleshooting, by symptom diagnosis, is based on diagnosing malfunctions, with possible causes arranged into categories relating to the operation of the refrigerator/freezer. This section is intended only to serve as a checklist to aid you in diagnosing a problem. Look at the symptom that best describes the problem you are experiencing with the refrigerator/freezer, then correct the problem.

Compressor will not run

1. Is there voltage at the wall receptacle? Check this with voltmeter.
2. Check for loose electrical connections.
3. Is the condenser coil dirty? A dirty condenser coil will overheat the compressor.

---

[4] Reprinted from MACAP Consumer Bulletin, issue no. 6, February 1984

4. Check the condenser fan motor.
5. Test the cold control for continuity.
6. Test the compressor, the relay, and the overload switch.

Compressor kicks out on overload

1. Check for high or low voltage when the compressor tries to start. High voltage will overheat the compressor. Low voltage will try to run the compressor with the start winding. A compressor is designed to start and run within a 10% tolerance of the rated voltage.
2. Test the capacitor. A shorted or open capacitor will overheat the compressor.
3. Test the compressor relay.
4. Test the overload for continuity.
5. Test the compressor windings for a short.

Refrigerator too cold

1. Check the damper control setting. Check to see if the damper is stuck open (thermostatically controlled dampers only).
2. Test the cold control switch contacts for continuity. Test for stuck contacts.
3. Check the location of the refrigerator. If outside in the winter, the ambient temperature may be too cold.

Refrigerator too warm

1. Check for restricted air circulation around the condenser coil.
2. Check the location of the refrigerator.
3. Check the door gaskets for proper sealing.
4. Check to see if the cabinet light is staying on when the door is closed.
5. Check the defrost heaters. Use a clamp-on ammeter (or wattmeter) to test the heaters, if they are coming on when the refrigeration cycle is running.
6. Check the cold control setting.
7. Check the compressor. Is it operating properly?
8. Is the evaporator fan running?
9. Check the air duct for restriction.
10. Check for a leaking air duct.
11. Check the evaporator coil for excessive frost build-up.
12. Check the defrost cycle. Is it working properly?
13. Check the damper control setting. Check to see if the damper is stuck closed (thermostatically controlled dampers only).

Refrigerator/freezer too noisy

1. Check for loose parts.
2. Check for rattling pipes.
3. Check the fan assembly, the evaporator, and the condenser.
4. Check the compressor.

5. If normal operational noises, so instruct the consumer.
6. Refrigerator/freezer not properly leveled.
7. Check the floor, it may not be structurally sound.

Sweating on the outside of the cabinet

1. Check the location of refrigerator/freezer. If located in an area of high humidity, it will begin to sweat.
2. Check for a void in the insulation between the cabinet and the inner liner.
3. Test the mullion and/or stile heaters for continuity.
4. Is the energy saver switch in the on position?
5. On older models, check for wet insulation.
6. Check for suction line, or any low side tubing, touching the cabinet.
7. Check for water leaks from icemaker.
8. Check chilled water supply lines and connections.
9. Check for a kinked, misaligned, or blocked drain system.
10. Check the defrost drain pan for misalignment, or for leaking cracks.
11. Are the doors aligned properly?

Sweating on the inside of the cabinet

1. Check for any abnormal usage. Instruct the consumer.
2. Check the door gaskets for proper sealing.
3. Check for defrost drain water leaking into the cabinet.
4. Is the condensate drain blocked?
5. Are the doors aligned properly?
6. Inspect all access holes where tubing or wires enter the refrigerator/freezer. Seal with perma-gum if necessary.
7. Inspect cabinet outer walls and seams for any openings. Seal with perma-gum if necessary.
8. Are there excessive door openings on hot, humid days?
9. Check for improper food storage.

Incomplete defrosting of the evaporator coil or high temperature during the defrost cycle

1. Test the defrost thermostat.
2. Check for loose wiring in the defrost electrical circuit.
3. Test the defrost timer for continuity.
4. Test for defective defrost heaters.

Odor in cabinet

1. Check for spoiling food in the cabinet.
2. Check the defrost water drain system.
3. Check the defrost heaters.

Excessive frost build-up on evaporator coil in freezer

1. Check the defrost cycle.
2. Check for loose wiring.
3. Is the heater making contact with the evaporator coil?
4. Check for proper door alignment.
5. Check the door gaskets.

Freezer run time too long

1. Check the thermostat setting.
2. Check for excessive loading of unfrozen food.
3. Check for incorrect wiring.

Temperature in freezer higher than normal

1. Check the thermostat for proper temperature calibration.
2. Test evaporator fan motor and blade.
3. Check the defrost timer.
4. Check for excessive loading of unfrozen food.
5. Check door gasket for proper sealing.

Refrigerator/freezer runs excessively or continuously

1. Check the interior lights for staying on continuously.
2. Check condenser coil for air restriction.
3. Check door gaskets.
4. On older models with automatic icemakers, check to be sure that the icemaker is operating properly.

SAVING THE OZONE LAYER

High above the earth is a layer of ozone gas that encircles the planet. The purpose of the gas is to block out most of the damaging ultraviolet rays from the sun. Such compounds as CFCs, HCFCs, and Halons have depleted the ozone layer, allowing more UV (ultraviolet) radiation to penetrate to the earth's surface.

In 1987, the United States, the European Economic Community, and 23 other nations signed the Montreal Protocol on Substances that Deplete the Ozone Layer. The purpose of this agreement was to reduce the use of CFCs throughout the world. To strengthen the original provisions of this protocol, 55 nations signed an agreement in London on June 29, 1990. At this second meeting, they passed amendments that called for a full phaseout of CFCs and Halons by the year 2000. Also passed at that meeting was the phaseout of HCFCs by the year 2020, if feasible, and no later than the year 2040, in any case.

On November 15, 1990, President Bush signed the 1990 Amendment to the Clean Air Act, which established a National Recycling and Emissions Reduction Program. This program minimizes the use of CFCs, and other substances harmful to the environment, while calling for the capture and recycling of these substances. The provisions of the Clean Air Act, signed by the president, are more stringent than those contained in the Montreal Protocol as revised in 1990.

Beginning July 1, 1992, the Environmental Protection Agency (EPA) developed regulations under Section 608 of the Clean Air Act (the Act), that limit emissions of ozone-depleting compounds. Some of these compounds are known as *chlorofluorocarbons (CFCs)* and *hydrochlorocarbons (HCFCs)*. The Act also prohibits releasing refrigerant into the atmosphere while maintaining, servicing, repairing, or disposing of refrigeration and air conditioning equipment. These regulations also require technician certification programs. A sales restriction on refrigerant is also included, whereby only certified technicians will legally be authorized to purchase such refrigerant. In addition, the penalties and fines for violating these regulations can be rather severe.

TROUBLESHOOTING SEALED-SYSTEM PROBLEMS

If you suspect a sealed-system malfunction, be sure to check out all external factors first. These include:

1. Thermostats.
2. Compressor.
3. Relay and overload on the compressor.
4. Interior lights.
5. Evaporator and condenser fans.
6. Timers.
7. Refrigerator/freezer getting good air circulation.
8. Food loaded in the refrigerator/freezer properly.
9. Check if heat exchanger has separated.

After eliminating all of these externals, you will then systematically check the sealed system. This is accomplished by comparing the conditions found in a normal operating refrigerator/freezer. These conditions are:

1. Refrigerator/freezer storage temperature.
2. Wattage.
3. Condenser temperature.
4. Evaporator inlet sound (gurgle, hiss, etc.)
5. Evaporator frost pattern.
6. High side pressure[5]
7. Low side pressure[5]
8. Pressure equalization time.

One thing to keep in mind: no single indicator is conclusive proof that a particular sealed system problem exists. Rather, it is a combination of findings that must be used to definitely pinpoint the exact problem.

---

[5] If you open up the sealed system, you will void your warranty. The sealed system must be repaired by an authorized service company.

Refrigerant leak

To determine a refrigerant leak in the sealed system, the symptoms are:
1. Temperatures in the storage area are below normal.
2. The wattage and amperage are below normal, as indicated on the model/serial plate.
3. The temperature of the condenser coil will be cool to the touch; at the last pass, or even as far as midway through the coil.
4. At the evaporator coil you will hear a gurgling noise, a hissing noise, or possibly an intermittent hissing or gurgling noise.
5. When the evaporator coil cover is removed, the evaporator coil will show a receded frost pattern.
6. The high and low side pressure will be below normal.[5]
7. The pressure equalization time might be normal, or shorter than normal.

Overcharged refrigerator/freezer

If the sealed system is overcharged, the symptoms are:
1. The storage temperature will be higher than normal.
2. The wattage and amperage are above normal, as indicated on the model/serial plate.
3. The temperature of the condenser coil will be above normal.
4. At the evaporator coil, you will hear a constant gurgling noise. Generally, a higher sound level than normal.
5. When the evaporator coil cover is removed, the evaporator coil will show a full frost pattern. If you remove the back cover, located behind the refrigerator/freezer, you will possibly see the suction line frosted back to the compressor.
6. The high and low side pressure will be above normal.[5]
7. The pressure equalization time will be normal.

Slight restriction

If you suspect a slight restriction in the sealed system, the symptoms are:
1. The storage temperature will be below normal.
2. The wattage and amperage are below normal, as indicated on the model/serial plate.
3. The temperature of the condenser coil will be slightly below normal.
4. At the evaporator coil, you will hear a constant gurgling noise and a low sound level.
5. When the evaporator coil cover is removed, the evaporator coil pattern will be receded.
6. The high and low side pressure will be below normal.[5]
7. The pressure equalization time will be longer than normal.

Partial restriction

If you suspect a partial restriction in the sealed system, the symptoms are:
1. The storage temperature will be higher than normal.
2. The wattage and amperage are below normal, as indicated on the model/serial plate.

3. The temperature of the condenser coil will be below normal, more than half way on the coil.
4. At the evaporator coil you will hear a constant gurgling noise, and a considerably low sound level.
5. When the evaporator coil cover is removed, the evaporator coil will be considerably receded.
6. The high and low side pressure will be below normal.[5]
7. The pressure equalization time will be longer than normal.

Complete restriction

If you suspect a complete restriction in the sealed system, the symptoms are:
1. The storage temperature will be warm.
2. The wattage and amperage are considerably below normal, as indicated on the model/serial plate.
3. The temperature of the condenser coil will be cool, or at room temperature.
4. At the evaporator coil, you will hear no sounds.
5. When the evaporator coil cover is removed, the evaporator coil will not have any frost on it, or the frost will be melting.
6. The high side pressure will be equal to the pressure of refrigerant at room temperature.
7. The low side pressure will be in a deep vacuum.[5]
8. There will be no pressure equalization time.

Moisture restriction

If you suspect a moisture restriction in the sealed system, the symptoms are:
1. The storage temperature will be above normal.
2. The wattage and amperage are considerably below normal, as indicated on the model/serial plate.
3. The temperature of the condenser coil will be below normal.
4. At the evaporator coil you will hear a constant gurgle, low sound level, or no sound at all.
5. When the evaporator coil cover is removed, the evaporator coil might have some frost on the evaporator inlet.
6. The high side pressure will be below normal.[5]
7. The low side pressure will be below normal or in a deep vacuum.[5]
8. The pressure equalization time will be longer than normal or there will be no equalization at all.

Low capacity compressor

To determine a low capacity compressor in the sealed system, the symptoms are:
1. Temperatures in the storage area are above normal.
2. The wattage and amperage are below normal, as indicated on the model/serial plate.
3. The temperature of the condenser coil will be below normal.
4. At the evaporator coil, you will hear a slightly reduced gurgling noise.

5. When the evaporator coil cover is removed, the evaporator coil will show a normal frost pattern.
6. The high side pressure will be below normal, and the low side pressure will be above normal.[5]
7. The pressure equalization time might be normal, or shorter than normal.

REPAIR PROCEDURES

Each repair procedure is a complete inspection and repair process for a single refrigerator/freezer component. It contains the information you need to test and replace components.

Door gasket

The typical complaints associated with door gasket failure are: 1. Sweating inside the cabinet. 2. Temperatures inside the cabinet are warmer than normal. 3. Ice forming on the freezer walls. 4. Door gaskets not maintained properly.

1. *Verify the complaint* Verify the complaint by checking the door gasket for proper sealing and alignment. Inspect the gaskets for any damage.
2. *Check for external factors* You must check for external factors not associated with the appliance. Is the appliance installed properly? Were the doors reinstalled correctly?
3. *Disconnect the electricity* Before working on the refrigerator/freezer, disconnect the electricity. This can be done by pulling the plug from the receptacle. Or disconnect the electricity at the fuse panel or at the circuit breaker panel. Turn off the electricity.
4. *Gain access to the door gaskets* To access the door gaskets, open the refrigerator/freezer door. The gaskets are located on the door.
5. *Test the door gaskets* To test the gaskets for proper sealing, take a dollar bill and place it between the gasket and the flange of the outer cabinet (Fig. 13-4). Pull on the dollar bill. When pulling on the dollar bill, you should feel some tension as it grips the bill. This means that the gasket is making good contact with the refrigerator/freezer flange. Repeat this test in other areas where you suspect problems with the gasket. If the gasket fails this test, then the next step is replace the gasket. For the doors to close and seal properly, the refrigerator should tilt backwards ¼ of an inch. This is accomplished by raising the front legs, or wheels, according to the installation instructions. If you are still unable to get the doors to close properly, then check the doors for sagging or warping (Fig. 13-5). Also, check the floor to see if it is level under the refrigerator/freezer. Check from front to back, and from side to side.
6. *Remove the door gasket* Before you get started, remove all of the food from the door. To remove the door gasket, pull back on the gasket, which exposes the retaining strip and screws. Loosen the screws about half way, but do not remove them (Fig. 13-6). Gently remove the gasket from around the door (Fig. 13-7).

13-4
The gasket must make full contact with the cabinet flange.

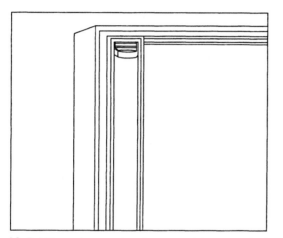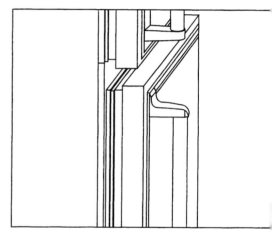

13-5 Warped or sagging doors must be corrected so that the refrigerator/freezer will operate properly.

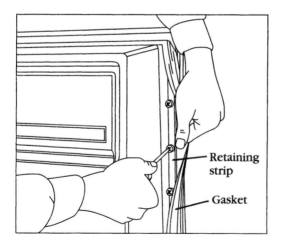

Retaining strip

Gasket

13-6
Peel back the gasket to gain access to the retaining strip and screws.

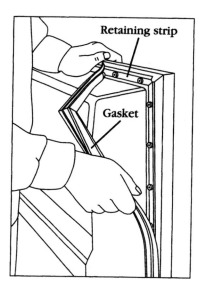

13-7
Carefully remove the gasket so as to not damage the inner door liner.

7. *Install a new door gasket* Before installing the new gasket, soak it in warm water. This will make the gasket soft and easier to install. Starting at either top corner, insert the flange of the gasket behind the retaining strip and/or door liner. Proceed all the way around the door. When the gasket is in place, begin to tighten the screws slightly all around the door. Now close the door, the gasket should make contact with the cabinet flange; evenly, and all around the door (Fig. 13-8). The door gasket might be adjusted by aligning the door panel, as in Fig. 13-9. To align the door, twist the door in the opposite direction of the warp. Close the door, and check for the gasket sealing against the cabinet. Now that the door gasket is sealing properly,

13-8
Have someone help you hold the door straight when tightening the screws.

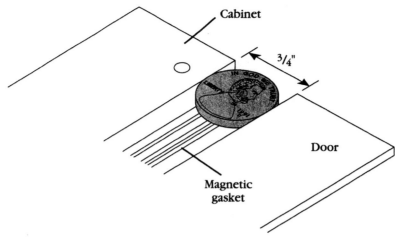

13-9 The correct gap between the door and cabinet will allow the door to close and seal properly. The gap should also be checked as the gasket starts rolling inward when the door is closed.

tighten the screws completely. If the gasket is distorted or if it has wrinkles in it, use a hair dryer to heat the gasket and mold it to its original form. Then, recheck to be sure the gasket seats against the flange properly. Next, check the gap between the door and the cabinet (Fig. 13-9) on the hinge side. Use a penny which is about ¾ of an inch in diameter to check the gap. Slide the penny from the top hinge to the bottom hinge. The door might be adjusted by moving the top hinge and by adding or removing shims to the center and bottom hinges. Be sure that the doors line up evenly with the sides of the cabinet and evenly with each other. Also, check that when both doors are opened simultaneously (top mount refrigerators only), that they do not hit against each other.

Thermostat (cold control)

The typical complaints associated with thermostat (cold control) failure are: 1. The refrigerator/freezer is not cold enough. 2. The refrigerator/freezer is too cold. 3. The refrigerator/freezer runs all the time. 4. The refrigerator/freezer doesn't run at all.

1. *Verify the complaint* Verify the complaint by checking the control setting. Turn the control off; then turn it on again, to see if the refrigerator/freezer starts.

2. *Check for external factors* You must check for external factors not associated with the appliance. Is the appliance installed properly? Explain to the user how to set the controls.

3. *Disconnect the electricity* Before working on the refrigerator/freezer, disconnect the electricity to the refrigerator/freezer. This can be done by pulling the plug from the receptacle. Or disconnect the electricity at the fuse panel or at the circuit breaker panel. Turn off the electricity.

4. *Gain access to the thermostat* To access the thermostat, open the fresh food door. Look for the control dial that has the word "off" printed on it. This is the control that turns the compressor on and off. Remove the dial (Fig. 13-10). Next, remove the two screws that secure the control (Fig. 13-11). Remove the wires from the terminals. On some models, the capillary tube is inserted in the air duct; but on other models, the capillary tube might be attached to the evaporator coil. At this point, if the capillary tube is attached to the coil, do not remove the capillary tube yet. Test the control first.

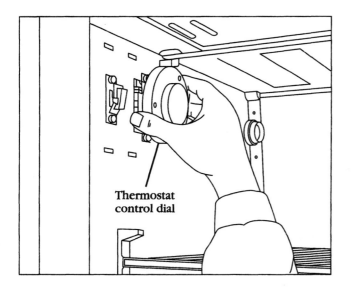

Thermostat control dial

13-10
Pull the dial off gently to gain access to the thermostat.

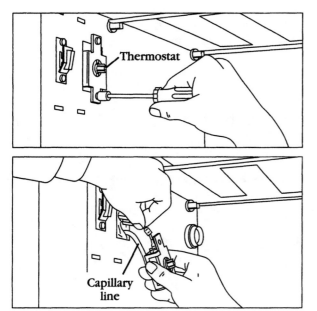

Thermostat

Capillary line

13-11
Be sure that the electric is disconnected before attempting to remove the thermostat. Note how the capillary tube is routed, and if it is secured to anything. Also, note in which direction the terminal end of the thermostat is pointed, when removing the thermostat. If by error, you install the thermostat upside down, the dial indicators will be incorrect. The thermostat inside the cabinet could be set on the wrong position.

5. *Test the thermostat* To test the thermostat, place the ohmmeter probes on the terminals (Fig. 13-12). Set the range scale on R × 1, and test it for continuity. With the control set in the off position, you should not read continuity. When the control is set to the highest position, you should read continuity. If the thermostat is good, the problem must be elsewhere.

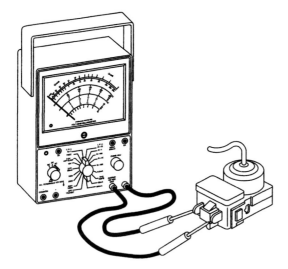

13-12
Check thermostat switch contacts for continuity. Also, inspect the capillary tube for any damage. If the capillary tube has lost its charge, the refrigerator/freezer might not cool or it could freeze the food in the fresh food compartment or keep the temperatures warmer than normal.

6. *Remove the thermostat* With the thermostat control housing already removed, the capillary tube must now be removed. Because there are many different models on the market today, Fig. 13-13 represents only a few types. The capillary tube might be routed through the control housing (Fig. 13-13A). The capillary tube might be secured to the evaporator by means of a clamp (Fig. 13-13B). The capillary tube might be inserted into a housing that senses how cold the air is (Fig. 13-13C). Whichever way the capillary tube is installed, remove it carefully so as not to damage the other components.
7. *Install a new thermostat.* To install the new thermostat just reverse the order of disassembly, and reassemble. Then, test the control. Just remember to reinstall the capillary tube in the same location from where it was removed. If you do not, the refrigerator/freezer will not work properly.

Defrost timer

The purpose of the defrost timer is to regulate the frequency of the defrost cycles, and the duration of each cycle. The typical complaints associated with defrost timer failure are: 1. The refrigerator/freezer does not defrost. 2. The storage temperature in the cabinet is too warm. 3. The compressor will not run.
1. *Verify the complaint.* Verify the complaint by asking the customer to describe what the refrigerator/freezer is doing or did.
2. *Check for external factors.* You must check for external factors not associated with the appliance. Is the appliance installed properly? Are the doors aligned properly?

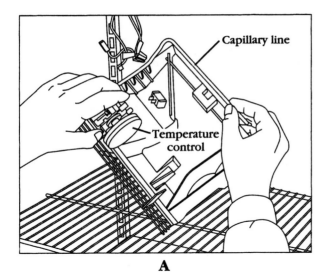

Capillary line

Temperature
control

A

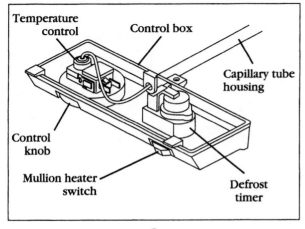

Evaporator

Thermostat
routing   Clamp

B

13-13
A. The capillary tube is attached to the control panel. It senses the air flow temperature. B. The capillary tube is attached to the evaporator plate. Note exactly where the capillary tube is attached. If you remove the capillary tube for any reason, you must reattach it back in the same location. C. Capillary tube location within the housing.

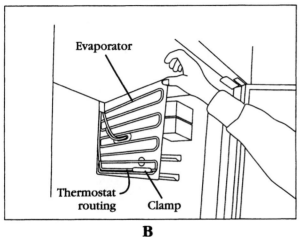

Temperature
control   Control box

Capillary tube
housing

Control
knob

Mullion heater
switch

Defrost
timer

C

3. *Disconnect the electricity.* Before working on the refrigerator/freezer, disconnect the electricity to the refrigerator/freezer. This can be done by pulling the plug from the receptacle. Or disconnect the electricity at the fuse panel or at the circuit breaker panel. Turn off the electricity.

4. *Gain access to the defrost timer.* To access the defrost timer, you must first locate it. On some models, the defrost timer is located on the bottom, behind the toe plate; or it might be behind the temperature control housing, in the fresh food section (Fig. 13-14); or it might be in the back of the refrigerator, behind the rear leg (Fig. 13-15).

5. *Remove defrost timer.* In order to test the defrost timer, it must be removed from its mounting position. Remove the two mounting screws from the defrost timer (Fig. 13-15). Next, remove the wire harness plug from the defrost timer (Fig. 13-16).

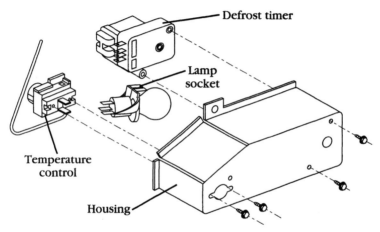

13-14 Defrost timer and thermostat located together in one housing.

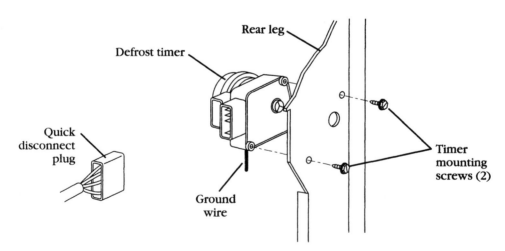

13-15 Remember to always reconnect the green ground wire to the defrost timer and ground.

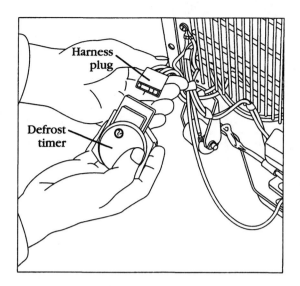

13-16
Not all defrost timers have harness plug connectors. Some are wired with individual wires. If you forget how to reconnect the wires properly, read the wiring diagram.

6. *Test the defrost timer.* To test the defrost timer, place a screwdriver in the timer cam slot (Fig. 13-17), and turn it clockwise until you hear the first "snap." The defrost timer is now in the defrost cycle. At this point in time, you must read the wiring diagram to determine which numbered terminals are for the defrost circuit. For the purpose of demonstrating how to check for continuity of the switch contacts, Fig. 13-18A illustrates the internal components of this sample timer. Set the ohmmeter scale on R × 1, and place the probes on the terminals marked 2 and 3 (Fig. 13-19). You should measure continuity. Next, rotate the timer cam until you hear the second "snap." The meter will show no continuity, indicating that the defrost cycle is over and that the refrigeration cycle begins.

Now, place the meter probes on the terminals marked 3 and 4. The ohmmeter will show continuity, indicating the refrigeration cycle is activated. Turn the timer cam once again, until you hear the first "snap." The meter will show no continuity. At no time should there be continuity

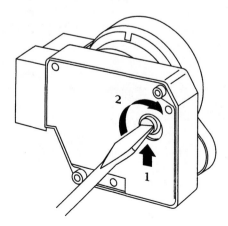

13-17
Always rotate the timer cam clockwise.

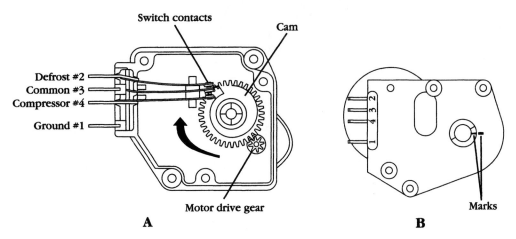

A

B

13-18 If you have continuity between terminal 2, 3, and 4, this indicates the switch contact points are all welded together. Symptoms in the refrigerator/freezer will be warmer temperatures than normal.

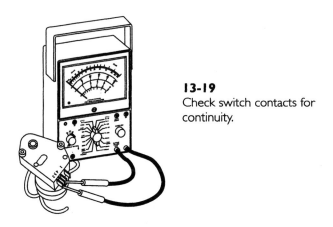

13-19
Check switch contacts for continuity.

between terminals 2 and 4. (If so, the switch contacts are burned and welded together. The defrost timer must be replaced.) If the defrost timer passes this portion of the test, then you must determine if the timer mechanism is functioning. Place the ohmmeter probes on the timer motor leads and read the resistance. The resistance can be between 800 and 4000 ohms, depending on which type of timer is used by the manufacturer. If you are unable to read resistance, then the timer motor is defective.

If the defrost timer passes this portion of the test, then rotate the timer cam until you hear the first "snap." Advance the timer cam again, counting the number of clicks until you hear the second "snap." Write down the number of clicks on a piece of paper. Now rotate the timer cam again until the marks line up (Fig. 13-18B), which indicates the beginning of the defrost cycle and the "snap" is heard. Advance the timer cam and count the clicks until there is one click left before the end of the defrost cycle. Take the timer and reconnect it to the wiring harness (Fig. 13-16). Place the defrost timer on a nonmetallic surface.

Be cautious when working with live wires. Avoid getting shocked.

Reconnect the voltage supply to the refrigerator/freezer. Place the ammeter jaws around the wire attached to the number four terminal. The meter should indicate no amperage. Next, place the jaws on the number two terminal wire. The ammeter should indicate some amperage. Wait for approximately 10 to 15 minutes, you should hear a "snap" indicating that the timer has completed the defrost cycle. At this point, the ammeter will show no amperage on number two, but will indicate current flow at number four. If not, replace the timer.

7. *Install a new defrost timer* To install the new defrost timer, just reverse the order of disassembly, and resemble. Remember to reconnect the ground wire to the defrost timer.

Evaporator fan motor

The typical complaints associated with evaporator fan motor failure are: 1. The refrigerator/freezer temperature is warm. 2. The evaporator fan motor runs slower than normal. 3. The evaporator fan motor does not run at all. 4. The evaporator fan motor is noisy.

1. *Verify the complaint* Verify the complaint by asking the customer to describe what the refrigerator/freezer is doing, or did. Is the evaporator fan motor running? Is it noisy?

2. *Check for external factors* You must check for external factors not associated with the appliance. Is the appliance installed properly? Is there something hitting the fan blade?

3. *Disconnect the electricity* Before working on the refrigerator/freezer, disconnect the electricity. This can be done by pulling the plug from the receptacle. Or disconnect the electricity at the fuse panel or at the circuit breaker panel. Turn off the electricity.

4. *Gain access to the evaporator fan motor* To access the evaporator fan motor, the evaporator cover must be removed (Fig. 13-20). Remove the screws that secure the cover in place. On some models, the evaporator fan assembly is located on the rear wall of the interior freezer compartment.

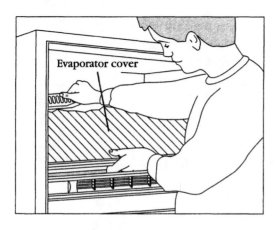

Evaporator cover

13-20
After removing the evaporator cover in this type of refrigerator, remove the heat shield to gain access to the components.

5. *Test the evaporator fan motor* The evaporator fan motor should be tested for proper resistance, as indicated on the wiring diagram. To test the evaporator fan motor, remove the wires from the motor terminals. Next, place the probes of the ohmmeter on the motor terminals (Fig. 13-21). Set the scale on R × 1. The meter should show resistance. If no reading is indicated, replace the motor. If the fan blade does not spin freely, replace the motor.

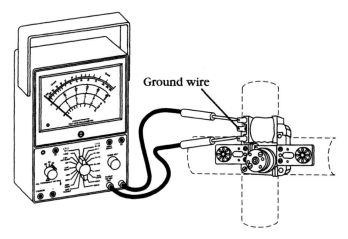

Ground wire

13-21
Check the evaporator fan motor for resistance. Also check the motor for grounded windings.

6. *Remove the evaporator fan motor* To remove the evaporator fan motor you must first remove the fan blade. On most models, just pull the blade off the motor shaft. Be careful not to break the blade. On other models, the fan blade is held on the motor shaft with screws. Remove the screws. Then, remove the screws that secure the fan assembly to the cabinet (Fig. 13-22). On some models, you must remove the fan shroud (Fig. 13-23), by removing the shroud screws.

7. *Install a new evaporator fan motor* To install the new evaporator fan motor, just reverse the order of disassembly, and reassemble. When reinstalling the fan blades onto the motor shaft, the fan blades should be positioned on the shaft so that one-third of its depth (approximately ¼") protrudes through the fan orifice, in the direction of air flow. When re-installing any shrouds, grilles, ducts, or gaskets, always position them correctly to ensure the proper air flow through the evaporator, and within both compartments of the refrigerator/freezer. Remember to reconnect the ground wire to the motor. Reconnect the wires to the motor terminals, and test.

Condenser fan motor

The typical complaints associated with condenser fan motor failure are: 1. The refrigerator/freezer temperature is warm. 2. The condenser fan motor runs slower than normal. 3. The condenser fan motor does not run at all. 4. The compressor is sometimes noisier than normal.

1. *Verify the complaint* Verify the complaint by asking the customer to describe what the refrigerator/freezer is doing. Is the condenser fan motor running?

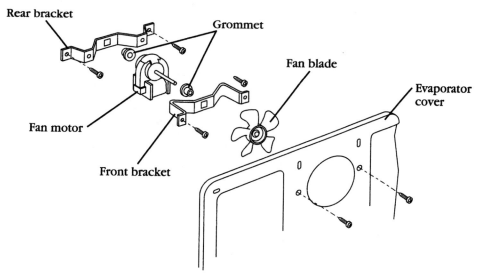

13-22 The exploded view of a evaporator fan motor assembly.

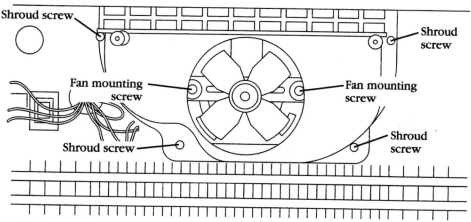

13-23 Removing the fan shroud to gain access to the motor terminals.

2. *Check for external factors* You must check for external factors not associated with the appliance. Is the appliance installed properly? Is there any foreign object blocking the condenser fan blade?

3. *Disconnect the electricity* Before working on the refrigerator/freezer, disconnect the electricity. This can be done by pulling the plug from the receptacle. Or disconnect the electricity at the fuse panel or at the circuit breaker panel. Turn off the electricity.

4. *Gain access to the condenser fan motor* Pull the refrigerator/freezer out and away from the wall. Remove the back panel, which is located at the bottom of the refrigerator/freezer. This will expose the compressor, the condenser fan assembly, and the condenser coil (Fig. 13-24).

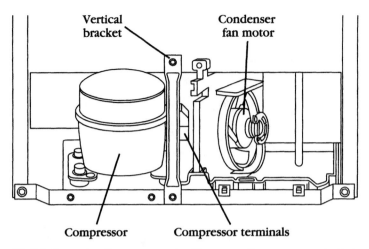

Vertical bracket

Condenser fan motor

Compressor **Compressor terminals**

13-24 Removing the rear panel will expose the components that need to be serviced.

5. *Test the condenser fan motor* A condenser fan motor should be tested for proper resistance, as indicated on the wiring diagram. Check the fan blade for obstructions. The blade should turn freely. Next, rotate the fan blade, and check for bad bearings. If you hear any unusual noises coming from the motor or if the fan blade is sluggish when spinning, replace the motor. To test the condenser fan motor, remove the wires from the motor terminals. Next, place the probes of the ohmmeter on the motor terminals (Fig. 13-25). Set the scale on R × 1. The meter should show resistance. If no reading is indicated, replace the motor.

6. *Remove the condenser fan motor* To remove the condenser fan motor, you must first remove the fan blades. Unscrew the nut that secures the blades to the motor. Remove the blades from the motor. Then, remove the motor assembly, by removing the mounting bracket screws (Fig. 13-26).

7. *Install a new condenser fan motor* To install the new condenser fan motor, just reverse the order of disassembly, and reassemble. Remember to reconnect the ground wire to the motor. Reconnect the wires to the motor terminals and test.

Defrost heater

The typical complaints associated with defrost heater failure are: 1. The refrigerator temperature is warm. 2. The freezer temperature is warm. 3. The refrigerator/freezer does not defrost. 4. Food loss. 5. No ice cubes.

1. *Verify the complaint* Verify the complaint by asking the customer to describe what the refrigerator/freezer is doing. Is there any food loss? Check the temperature in the compartments. Check for ice build-up on the evaporator cover.

2. *Check for external factors* You must check for external factors not associated with the appliance. Is the appliance installed properly?

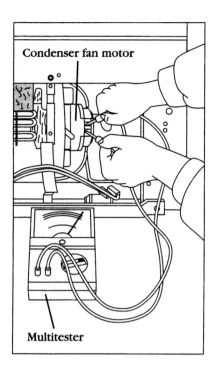

13-25
Remember to set the meter on the ohm scale when testing for resistance in the condenser fan motor.

Condenser fan motor

Multitester

3. *Disconnect the electricity* Before working on the refrigerator/freezer, disconnect the electricity. This can be done by pulling the plug from the receptacle. Or disconnect the electricity at the fuse panel or at the circuit breaker panel. Turn off the electricity.

4. *Gain access to the defrost heater* To access the defrost heater, the evaporator cover must be removed (Fig. 13-20). Remove the screws that secure the cover in place.

5. *Test the defrost heater* A defrost heater should be tested for proper resistance, as indicated on the wiring diagram. To test the defrost heater, remove the wires from the heater terminals. Next, place the probes of the ohmmeter on the heater terminals (Fig. 13-27). Set the scale on R × 1. The meter should show resistance. If no reading is indicated, replace the defrost heater.

6. *Remove the defrost heater* To remove the defrost heater in this type of model, you must first remove the reflector shield (Fig. 13-28). Bend the clip up and lift the shield. Do the same for the other end of the heater. Once the shield is removed, you can lift the defrost heater from its brackets.

7. *Install a new defrost heater* To install the new defrost heater (Fig. 13-29), just reverse the order of disassembly, and reassemble. One important note to remember: on some models, do not touch the glass because it will shorten the life of the heater. Remember to reconnect the wires to the heater. When reinstalling any shrouds, grilles, ducts, or gaskets, always position them correctly to ensure the proper air flow through the evaporator and within both compartments of the refrigerator/freezer.

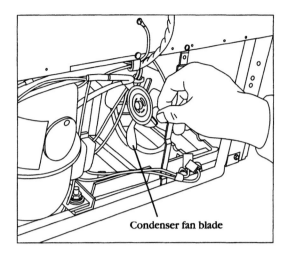

Condenser fan blade

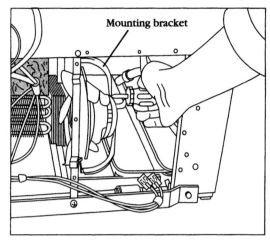

Mounting bracket

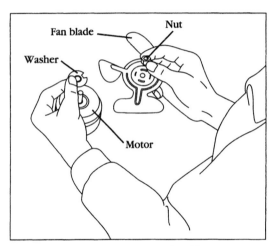

Fan blade

Nut

Washer

Motor

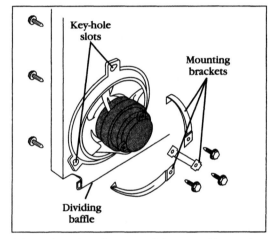

Key-hole slots

Mounting brackets

Dividing baffle

13-26 Removing the fan blade first. This will prevent the blades from bending out of shape, and becoming off-balanced. When removing the condenser fan motor, on some models, the compressor is located within inches of the motor. *Be careful!* The compressor might be hot and you could burn yourself.

Defrost thermostat

The typical complaints associated with defrost thermostat failure are: 1. The refrigerator temperature is warm. 2. The freezer temperature is warm. 3. The refrigerator/freezer does not defrost. 4. Food loss. 5. No ice cubes.

    1. *Verify the complaint* Verify the complaint by asking the customer to describe what the refrigerator/freezer is doing. Is there any food loss? Check the temperature in the compartments. Check for ice build-up on the evaporator cover.

    2. *Check for external factors* You must check for external factors not associated with the appliance. Is the appliance installed properly?

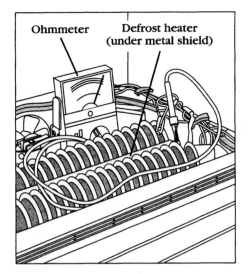

13-27
If the model that you are repairing has glass defrost heaters in it that look black in color, or dark smokey-grey, or look burned; then, the heater is defective, or soon will be.

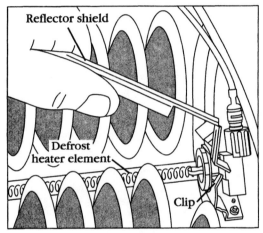

13-28
When removing this type of heater, be careful not to break the glass. The clip must be straight so the heater can slide out of the brackets.

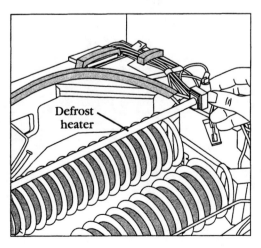

13-29
After the heater is installed and the shield is back in place, remember to bend back the clip on the bracket. When you reinstall the wires on the terminals, use one hand to hold the end of the heater. This will prevent cracking the glass if you push down too hard.

3. *Disconnect the electricity* Before working on the refrigerator/freezer, disconnect the electricity. This can be done by pulling the plug from the receptacle. Or disconnect the electricity at the fuse panel or at the circuit breaker panel. Turn off the electricity.

4. *Gain access to the defrost thermostat* To access the defrost thermostat, the evaporator cover must be removed (Fig. 13-20). Remove the screws that secure the cover in place.

5. *Test the defrost thermostat* A defrost thermostat failure usually results in a frost blocked evaporator. To test the defrost thermostat, disconnect the wires to isolate the thermostat from the rest of the defrost circuit. Next, place the probes of the ohmmeter on the defrost thermostat wire leads (Fig. 13-30). Set the meter scale on R × 1. The meter will show continuity when the thermostat is frozen, or very cold, indicating the defrost thermostat is good. The defrost thermostat switch contacts close when the temperature is colder than its temperature rating (Fig. 13-31). If no reading is indicated, replace the defrost thermostat. At ambient temperature, you will read no continuity, which will indicate the thermostat might be good.

6. *Remove the defrost thermostat* To remove the defrost thermostat you must remove the hold down clamp. On some models, the defrost thermostat and

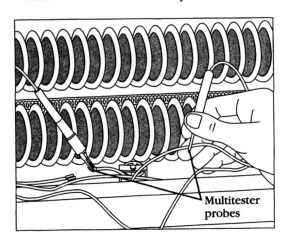

Multitester probes

13-30
Test the defrost thermostat when it is connected to the evaporator coil.

Part number	Temperature setting °F	
	Open	Close
L-45	45	25
L-50	50	30
L-55	55	35
L-60	60	40
L-70	70	50
L-80	80	30
L-90	90	35

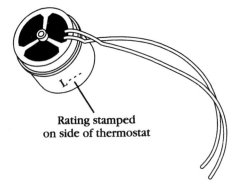

Rating stamped on side of thermostat

13-31 The temperature rating chart for most common defrost thermostats.

clamp are one assembly. Also, on some models, the defrost thermostat clamps around the evaporator tubing. In Fig. 13-32, remove this type of defrost thermostat by just squeezing in on the clip and lifting the thermostat up.

7. *Install a new defrost thermostat* To install the new defrost thermostat, just reverse the order of disassembly, and reassemble. Remember to reconnect the wires to the thermostat. When reinstalling any shrouds, grilles, ducts, or gaskets, always position them correctly to ensure the proper air flow through the evaporator and within both compartments of the refrigerator/freezer. On models that have the defrost thermostat attached to the evaporator coil, you must reinstall the defrost thermostat back in the same location from which it is removed.

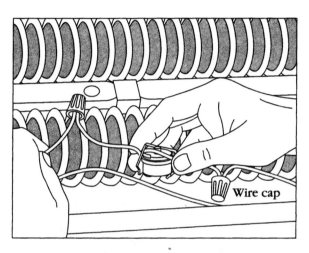

13-32
When replacing the defrost thermostat, be sure you reinstall it back in the same position from where you removed it. Otherwise, the defrost cycle will not function properly.

Compressor, relay, overload protector

The typical complaints associated with compressor failure are: 1. The refrigerator temperature is warm. 2. The freezer temperature is warm. 3. The refrigerator/freezer does not run at all. 4. Food loss.

1. *Verify the complaint* Verify the complaint by asking the customer to describe what the refrigerator/freezer is doing.

2. *Check for external factors* You must check for external factors not associated with the appliance. Is the appliance installed properly? Does the refrigerator/freezer have the correct voltage? Check for a voltage drop during refrigerator start-up.

3. *Disconnect the electricity* Before working on the refrigerator/freezer, disconnect the electricity. This can be done by pulling the plug from the receptacle. Or disconnect the electricity at the fuse panel or at the circuit breaker panel. Turn off the electricity.

4. *Gain access to the compressor* To access the compressor, pull the refrigerator/freezer out and away from the wall. Remove the back panel, which is located at the bottom of the refrigerator/freezer. This will expose the compressor, the condenser fan assembly, and the condenser coil (Fig. 13-24). Next, remove the compressor terminal cover (Fig. 13-33) by removing the retaining clip that secures the cover. Remove the terminal cover.

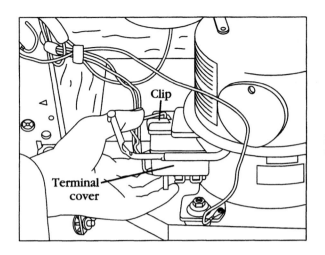

13-33
Removing the terminal cover to gain access to the relay and overload protector.

Terminal cover

Clip

5. *Test the compressor relay* To test the compressor relay, remove the relay by pulling it off of the compressor terminals, without twisting it (Fig. 13-34A). Remove the wires from the relay and label them. On the relay body is stamped the word TOP. Hold the relay so that TOP is in the up position. Next, place the probes of the ohmmeter on the relay terminals marked S and M. Set the meter scale on R × 1. The reading will show no continuity. Next, remove the probe from the terminal marked M, and place the probe on the side terminal marked L. The reading will show no continuity. Now, move the probe from terminal S, and place it on the terminal marked M. The reading will show continuity. With the probes still attached, turn the relay upside down (Fig. 13-34B) and perform the same tests. By turning the relay over, the switch contacts in the relay will close. When you retest the relay, you should get the opposite results. You should have continuity

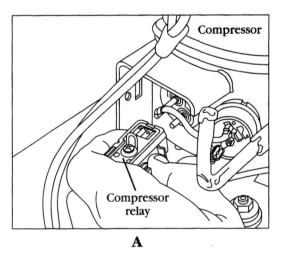

Compressor

Compressor relay

A

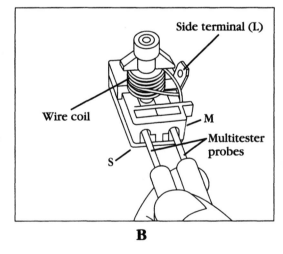

Side terminal (L)

Wire coil

M

Multitester probes

S

B

13-34 Pulling the relay off the terminals without twisting the relay. This will prevent you from breaking the compressor terminals.

between terminals S and M, and between S and L; however, the meter will not read continuity between M and L. If the relay fails this test, replace it.

6. *Test the overload protector* To test the overload protector, remove the wires from the overload and compressor terminals. Then, remove the overload protector from the compressor by removing the retaining clip that secures the overload protector to the compressor (Fig. 13-35A). Next, place the probes of the ohmmeter on the overload terminals (Fig. 13-35B). Set the meter scale on R × 1. The reading will show continuity. If not, replace the overload protector.

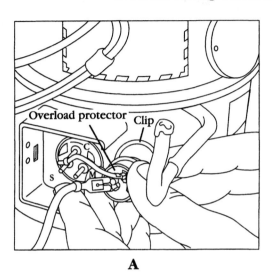

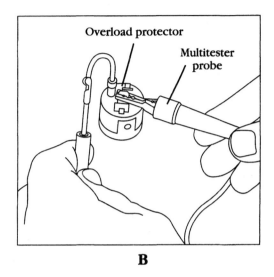

A   **B**

13-35 Testing the overload protector for continuity between the terminals.

7. *Test the compressor* To test the compressor, remove the relay and the overload protector. This will expose the compressor terminals. The compressor terminals are marked C, S, and R. The C indicates the common winding terminal; the S indicates the start winding terminal; and the R indicates the run winding terminal. Refer to the wiring diagram for the model that you are servicing. Set the meter scale on R × 1; touch the probes together, and adjust the needle setting to indicate a 0 reading. Next, place the probes of the ohmmeter on the terminals marked S and R (Fig. 13-36A). The meter reading will show continuity. Now, place the meter probes on the terminals marked C and S. The meter reading will show continuity. Finally, place the meter probes on the terminals marked C and R. The meter reading will show continuity. The total number of ohms measured between S and R is equal to the sum of C to S plus C to R. The compressor should be tested for proper resistance, as indicated on the wiring diagram.

To test the compressor for ground, place one probe on a compressor terminal, and attach the other probe to the compressor housing, or any good ground (Fig. 13-36B). Set the meter scale to R × 1000. The meter reading will show no continuity. Repeat this for the remaining two terminals. The meter reading will show no continuity. If you get a continuity reading from any of these terminals to ground, the compressor is grounded. Replace it.

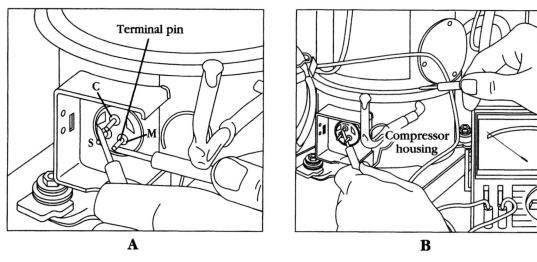

$$\text{A} \qquad\qquad\qquad\qquad \text{B}$$

13-36 Testing the compressor motor windings for resistance.

UPRIGHT FREEZERS

Upright freezers are similar to refrigerator/freezers in design and operation. They share some of the same features:

- Fan motors
- Compressor
- Automatic defrost system
- Door gasket
- Thermostat
- Interior lighting

On manual defrost models[7], the evaporator coils are the shelves inside the cabinet. Figure 13-37 illustrates the refrigerant flow on this type of manually defrosted upright freezer. The condenser coils are embedded between the cabinet liners and are secured to the inside wall of the outer cabinet. This provides for even heat removal and it eliminates the need for a condenser fan motor.

Automatic defrost models[8] use a fan motor to circulate the air inside the cabinet through air ducts. The evaporator coil is mounted on the inside back wall of the inner liner. Figure 13-38 illustrates the air flow pattern of this type of automatic-defrost, upright freezer. Figure 13-39 illustrates the refrigerant flow in this type of upright freezer.

To diagnose and repair the upright freezer, consult the troubleshooting and the repair procedures sections of this chapter.

---

[7] Manual-defrost freezers must be defrosted once or twice a year, depending upon the usage.

[8] Automatic-defrost freezers will self defrost the evaporator coil. The defrost components are the defrost timer, the defrost thermostat, and the defrost heater.

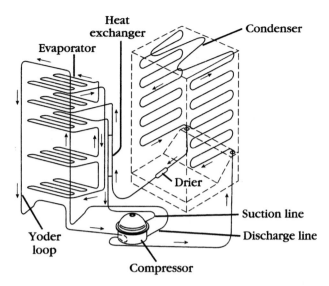

13-37 Refrigerant flow of an upright freezer with manual defrost.

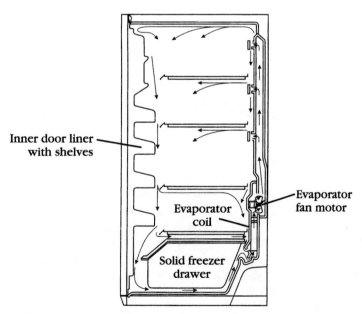

13-38 The air flow pattern of an automatic defrost upright freezer.

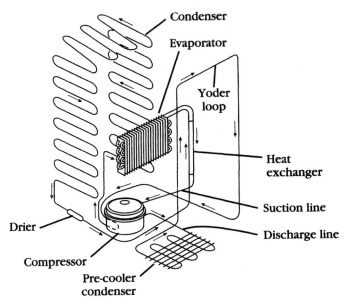

Condenser

Evaporator

Yoder
loop

Heat
exchanger

Suction line

Discharge line

Drier

Compressor

Pre-cooler
condenser

13-39 The refrigerant flow of upright freezer automatic defrost.

CHEST FREEZERS

The chest freezer (Fig. 13-40) has the evaporator coils and the condenser coils embedded between the inner liner and the outer cabinet. These coils are inaccessible for replacement or repair, if a refrigerant leak occurs. The differences between the upright freezer and the chest freezer are:

- Door hinges
- Gasket
- Location and access of temperature controls
- Location and access of the compressor, relay, and overload protector. Most models have a power indicator light. This light stays on as long as the freezer is plugged into the wall receptacle. The light alerts the consumer when the power is off to the freezer. But, it does not tell you what the temperature is inside the cabinet.

Chest freezers must be defrosted once or twice a year to remove the ice build-up from the inside. To gain access to the components, remove the side access panel (Fig. 13-41).

Repair procedures for chest freezers

Each repair procedure is a complete inspection and repair process for a single freezer component; containing the information you need to test a component that might be faulty; and to replace it, if necessary.

Thermostat (cold control)

The typical complaints associated with thermostat (cold control) failure are: 1. The freezer is not cold enough. 2. The freezer is too cold. 3. The freezer runs all the time. 4. The freezer doesn't run at all.

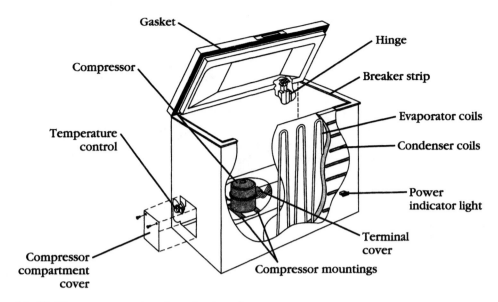

13-40 The component location of a chest freezer.

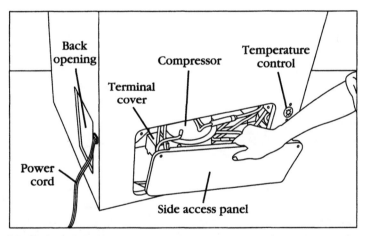

13-41 Removing side access panel to gain access to the components.

1. *Verify the complaint* Verify the complaint by checking the control setting. Turn the control off and on again to see if the freezer starts up. Is the power indicator light on?
2. *Check for external factors* You must check for external factors not associated with the appliance. Is the appliance installed properly? Explain to the user how to set the controls.
3. *Disconnect the electricity* Before working on the freezer, disconnect the electricity. This can be done by pulling the plug from the receptacle. Or disconnect the electricity at the fuse panel or at the circuit breaker panel. Turn off the electricity.

4. *Gain access to the thermostat.* To access the thermostat, remove the access panel (Fig. 13-41). Next, remove the two screws that secure the control. Remove the wires from the terminals. The capillary tube is inserted into a channel. Do not remove capillary tube yet. Test the control first.

5. *Test the thermostat* To test the thermostat, place the ohmmeter probes on the terminals (Fig. 13-42). Set the range scale on R × 1, and test for continuity. With the control set in the off position, you should not read continuity. When the control is set to the highest position, you should read continuity. If the thermostat is good, the problem must be elsewhere.

6. *Remove the thermostat.* With the thermostat control housing already removed, the capillary tube must now be removed. Remove the capillary tube from the channel.

7. *Install a new thermostat* To install the new thermostat just reverse the order of disassembly, and reassemble. Then, test the control. Remember to reinstall the capillary tube in the same location from where it was removed. Be careful as not to kink the tube. If you do, the freezer will not work properly.

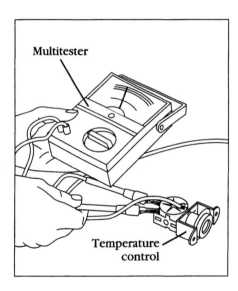

13-42
Testing the thermostat with a multitester, with the range set on the ohms scale. If you read no continuity, replace the thermostat.

Power indicator light

The typical complaints associated with the power indicator light failure are: 1. The light is not on. 2. The light flickers. 3. The light is dim.

1. *Verify the complaint* Verify the complaint by checking if the freezer is plugged in to the wall receptacle. Is the power indicator light on?

2. *Check for external factors* You must check for external factors not associated with the appliance. Is the appliance installed properly? Check the voltage to the freezer.

3. *Disconnect the electricity* Before working on the freezer, disconnect the electricity. This can be done by pulling the plug from the receptacle. Or disconnect the electricity at the fuse panel or at the circuit breaker panel. Turn off the electricity.

4. *Gain access to the power indicator light* To access the power indicator light, use a screwdriver to pry out the power indicator light from the front of the freezer cabinet. Next, remove the wires from the indicator light (Fig. 13-43).

5. *Test the power indicator light* To test the power indicator light, place the ohmmeter probes on its terminals (Fig. 13-44). Set the range scale on R × 1, and test for continuity. The meter should show continuity; if not, replace the component.

6. *Installing a power indicator light* To install the new power indicator light, just reverse the order of disassembly, and reassemble (Fig. 13-45). Then, test the control.

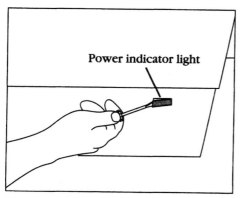

13-43 Pry gently to remove the power indicator light. Be careful not to scratch the cabinet.

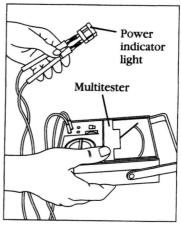

13-44 Testing the power indicator light.

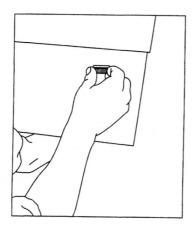

13-45
When installing the power indicator light, you must first reconnect the wires and then insert the light in the cabinet.

Gasket

The typical complaints associated with door gasket failure are: 1. Sweating inside and/or outside of the cabinet. 2. Temperatures inside the cabinet are warmer than normal. 3. Ice forming on the freezer walls. 4. Door gasket not maintained properly.

1. *Verify the complaint* Verify the complaint by checking the door gasket for proper sealing and alignment. Inspect the gaskets for any damage.
2. *Check for external factors* You must check for external factors not associated with the appliance. Is the appliance installed properly?
3. *Disconnect the electricity* Before working on the refrigerator/freezer, disconnect the electricity. This can be done by pulling the plug from the receptacle. Or disconnect the electricity at the fuse panel or at the circuit breaker panel. Turn off the electricity.
4. *Test the door gasket* To test the gasket for proper sealing, take a dollar bill and place it between the gasket and the flange of the outer cabinet. Pull on the dollar bill. When pulling on the dollar bill, you should feel some tension as the gasket and flange grip the bill. Repeat this test in other areas where you suspect problems with the gasket. If the gasket fails this test, the next step is to replace the gasket.
5. *Gain access to the door gasket* To access the door gasket, the door must be removed (Fig. 13-46). Turn the door over on its back.
6. *Remove the gasket.* Remove the gasket by prying the studs out, or by removing the screws (Fig. 13-47).
7. *Install a new door gasket* Before installing the new gasket, soak it in warm water. This will make the gasket soft and easier to install. Starting at either top corner, insert the flange of the gasket behind the retaining strip and/or door liner. Proceed all the way around the door. When the gasket is in place, begin to tighten the screws slightly all around the door, or reinstall the studs. If the gasket is distorted or if it has wrinkles in it, use a hair dryer to heat the gasket and mold it to its original form. Then, be sure the gasket seats against the flange properly. Next, check the gap between the door and the cabinet; adjust it, if necessary (Step 4).

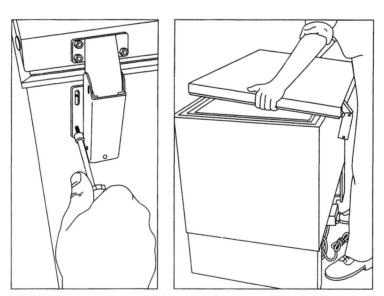

13-46 Removing the screws from the hinges and lifting the door from the freezer.

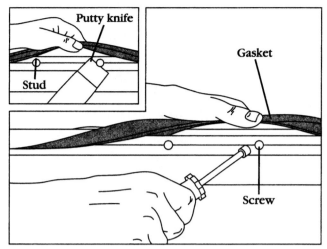

13-47 Removing the studs (inset). On some models, you must remove the screws that secure the gasket to the door.

DIAGNOSTIC CHARTS

The following refrigerator and freezer diagnostic charts will help you to pinpoint the likely causes of the particular problem (Figs. 13-48, 13-49, 13-50, and 13-51).

REFRIGERATOR/FREEZER ELECTRICAL SCHEMATICS

The wiring diagrams in this chapter are for examples only. You must refer to the wiring diagram on the refrigerator/freezer you are servicing. Figure 13-52 depicts an actual wiring schematic for a side-by-side refrigerator, which includes an automatic icemaker, and ice and water dispensers. On models with the rapid electrical diagnosis (R-E-D) feature, a technician can make a quick and accurate diagnosis of electrical faults without disassembling the refrigerator.[9] To perform this test, a special adapter is connected into the wiring harness through the multi-circuit connector, located behind the front grill. Upon separating the multi-circuit connector, the parallel circuits in the wiring harness will be isolated. This process will permit you to test all of the electrical components, and the related wiring within the main wiring harness. If a R-E-D test adapter is not available, you can still check the circuits with a ohmmeter.

Caution: Disconnect the electricity from the refrigerator/freezer before measuring resistances.

A basic understanding of the symbols used in the schematic diagram is essential (Fig. 13-52). The numbered terminals, located in the multi-circuit connector, are shown on the schematic diagram. The component circuits on the schematic diagram are indicated by an arrow and a number. The point of the arrow indicates a male terminal, and the tail of the arrow indicates a female terminal. The number identifies the terminal location in the connector (Fig. 13-53).

Remember how to read a wiring schematic. Give this a try.

---

[9] Robinair part # 14442, purchased at your local parts dealer.

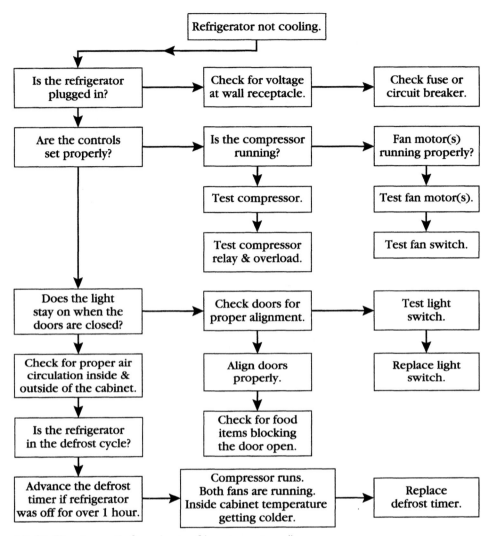

13-48 The diagnostic flow chart: refrigerator not cooling.

Example #1

The customer explains that the food in the freezer is thawing, and the food in the refrigerator is warm. The customer states the compressor runs constantly. You made your observation, and confirmed that the food is thawing and the compressor is operating. You also noticed frost on the evaporator cover, which indicates a defrost problem. This indicates to you the possibility of three components malfunctioning. These components are: the defrost heaters, the defrost thermostat, or the defrost timer. Also, there is a possibility of a broken or loose wiring connection. Set the refrigerator in the defrost mode first. Advance the defrost timer until you hear the first snap sound coming from the timer.

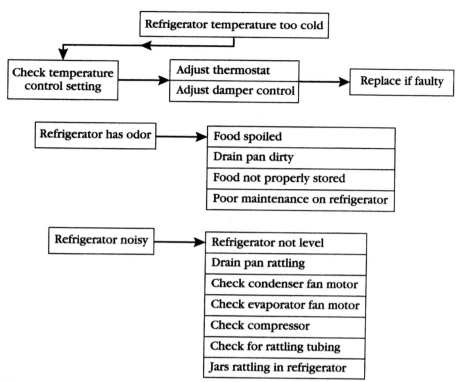

13-49 The diagnostic flow chart: refrigerator temperature too cold.

Using the ohmmeter, set the range on R × 1, and adjust the meter scale to read at zero. Separate the multicircuit connector. *Did you turn off the electricity?* To test the defrost heaters, insert one meter probe into the number "5" male connector pin, and the other meter probe into the number "1" male connector pin (Fig. 13-53A). The ohmmeter should show the combined resistance of the heaters. This resistance is 16 ohms. If the test passes, continue on to the defrost thermostat. If not, replace the heaters. When testing the defrost thermostat, insert one meter probe into the number "1" male connector pin, and the other meter probe onto the number "4" male connector pin (Fig. 13-53B). The meter reading should indicate "0" ohms. If not, replace the defrost thermostat. Now, test the defrost timer motor. Insert one meter probe onto the number "4" male connector pin, and the other meter probe into the number "4" female connector pin (Fig. 13-53C). The meter should show between 800 and 4000 ohms. Adjust the ohmmeter range accordingly. To test the defrost timer switch contacts, insert one meter probe into the number "2" female connector pin, and the other meter probe into the number "5" female connector pin (Fig. 13-53D). The meter should show 0 ohms. If not, replace the defrost timer control.

Figure 13-54 depicts the active defrost circuits. If you want to check the entire defrost circuit, insert one meter probe onto the number "3" male connector pin, and the other meter probe onto the number "4" male connector pin. The meter should read continuity. Be sure that the cold control (thermostat) is set for maximum cooling, and that the defrost timer set on defrost.

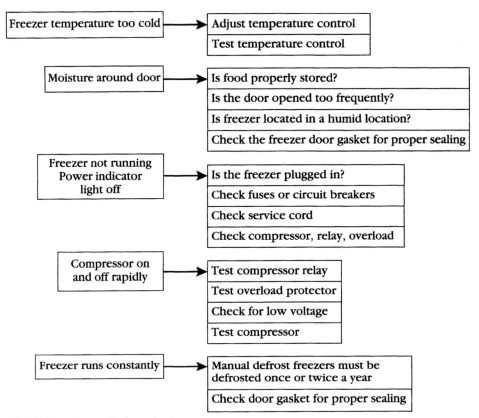

```
Freezer temperature too cold ──────► Adjust temperature control
                                      Test temperature control

      Moisture around door ──────►    Is food properly stored?
                                      Is the door opened too frequently?
                                      Is freezer located in a humid location?
                                      Check the freezer door gasket for proper sealing

       Freezer not running
        Power indicator ──────►       Is the freezer plugged in?
           light off                  Check fuses or circuit breakers
                                      Check service cord
                                      Check compressor, relay, overload

        Compressor on
       and off rapidly ──────►        Test compressor relay
                                      Test overload protector
                                      Check for low voltage
                                      Test compressor

   Freezer runs constantly ──────►    Manual defrost freezers must be
                                      defrosted once or twice a year
                                      Check door gasket for proper sealing
```

13-50 The diagnostic flow chart.

Example #2

The wiring diagrams in Figs. 13-55 and 13-57 are examples only. You must refer to the wiring diagram on the refrigerator that you are servicing. Figure 13-55 depicts an actual wiring schematic for a cycle-defrost refrigerator.

The refrigerator door is closed, the thermostat is calling for cooling, and the compressor is running. What circuits are active? Trace the active circuits in Fig. 13-55.

The active components are the thermostat, overload, relay, compressor, stile heater, and the mullion heater. Check the end result with Fig. 13-56.

The thermostat in Fig. 13-57 has cycled off, and the switch contacts are open. Also, the compressor stopped running. What circuits are active? Trace the active circuits in Fig. 13-57. The active components are the evaporator heater, drain heater, stile heater, and the mullion heater. Check the end result with Fig. 13-58. Current will flow through the overload, the relay coil, and the compressor running winding. But, there is not enough current to energize the start relay, and to run the compressor.

Example #3

The wiring diagram in Fig. 13-59 is an example only. It depicts a no-frost refrigerator. The refrigerator is in the cooling mode, and the compressor is running. What circuits are active? Trace the active circuits. Check the end result with Fig. 13-60.

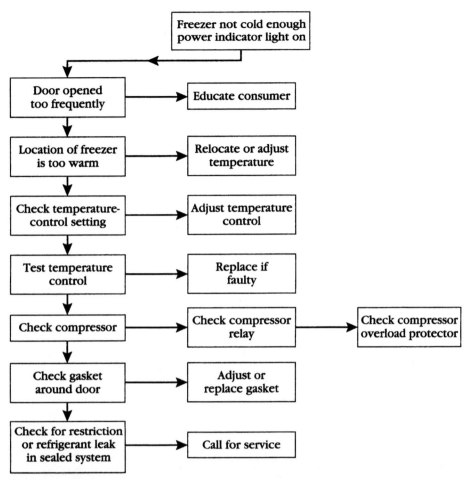

13-51 The diagnostic flow chart: freezer not cold enough, power indicator light on.

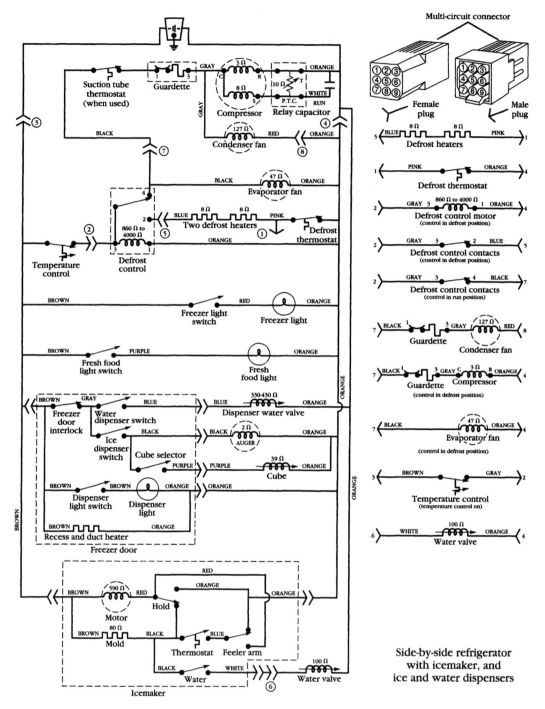

13-52 Side-by-side refrigerator wiring schematic.

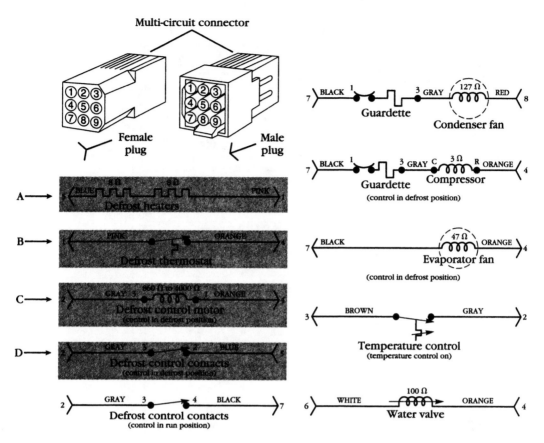

13-53 Side-by-side refrigerator wiring schematic of multi-circuit connector circuitry. A. Defrost heater circuit. B. Defrost thermostat circuit. C. Defrost control motor circuit. D. Defrost control switch contacts circuit.

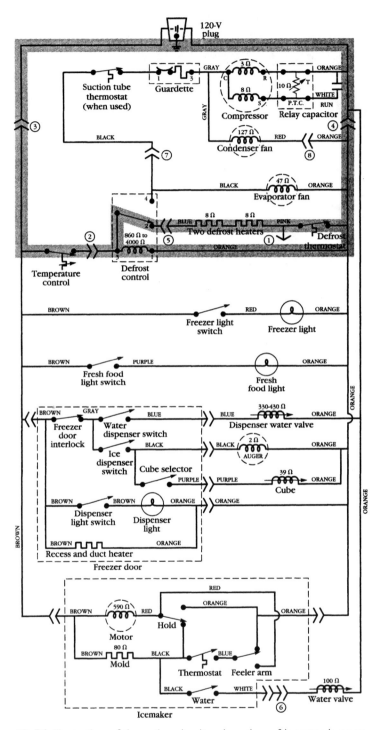

13-54 Illustration of the active circuits when the refrigerator is operating in the defrost mode.

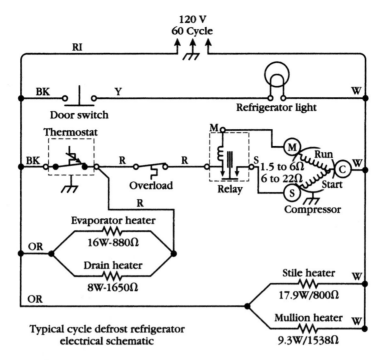

13-55 A typical cycle-defrost refrigerator electrical schematic.

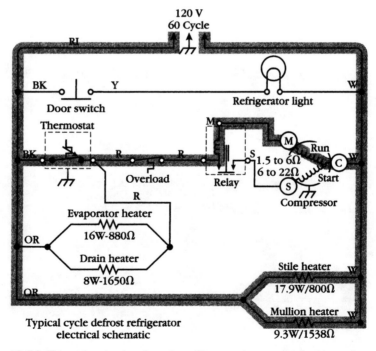

13-56 The active circuits when the refrigerator is operating in the cooling mode.

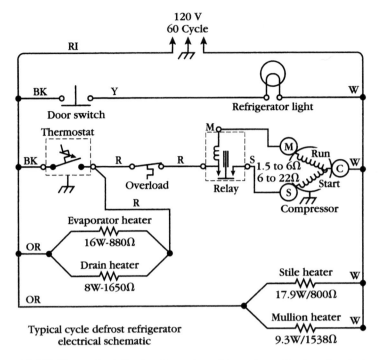

13-57 The cycle-defrost refrigerator wiring schematic, showing that the refrigerator has cycled off.

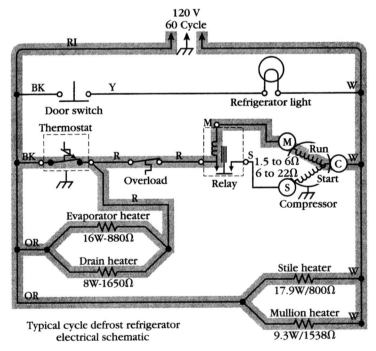

13-58 The active circuits when the refrigerator is operating in the defrost mode.

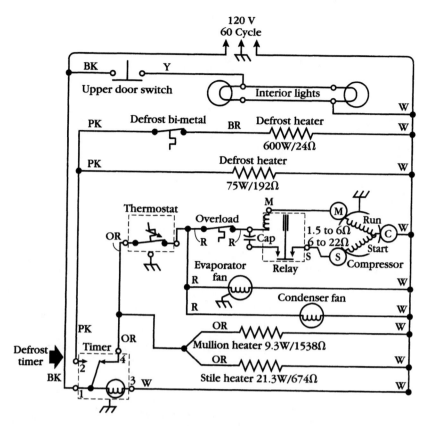

13-59 A typical no-frost refrigerator electrical schematic.

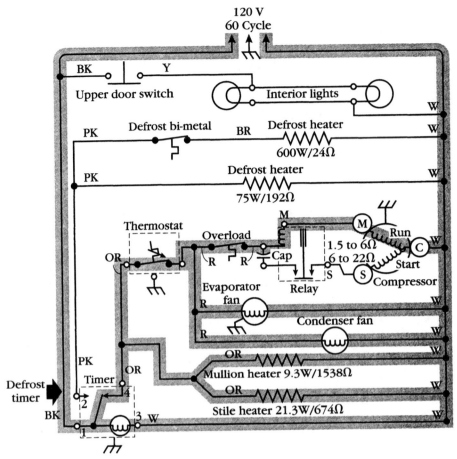

13-60 The active circuits when the refrigerator is operating in the cooling mode.

14
Automatic ice makers

The automatic ice maker is an independent appliance, installed in the freezer, or in the freezer compartment of a refrigerator. The ice maker has three distinct operations: fill, freeze and harvest the ice cubes. It is the refrigerator/freezer cooling system that allows the water to freeze into whatever shape is designed by the ice mold or tray. The amount of time it takes for the ice maker to produce and harvest ice cubes will depend on:

- Freezer temperature
- Food load conditions
- Amount of door openings
- Ambient temperature

SAFETY FIRST

Any person who cannot use basic tools, or follow written instructions, should *not* attempt to install, maintain, or repair any automatic ice makers. Any improper installation, preventive maintenance, or repairs could create a risk of personal injury or property damage.

If you do not fully understand the installation, preventive maintenance, or repair procedures in this chapter or if you doubt your ability to complete the task on your automatic ice maker, then please call your service manager.

The following precautions should also be followed:

1. Never place fingers or hands on the automatic ice maker mechanism while the refrigerator/freezer is plugged in.
2. Disconnect the electrical supply to the freezer or refrigerator before servicing the automatic ice maker.
3. Be careful of any sharp edges on the automatic ice maker, which might result in personal injury.
4. Do not attempt to operate the automatic ice maker unless it has been properly reinstalled, including the grounding and electrical connections.

Before continuing, take a moment to refresh your memory on safety procedures in Chapter 2.

AUTOMATIC ICE MAKERS IN GENERAL

Much of the troubleshooting information in this chapter covers automatic ice makers in general, rather than specific models, in order to present a broad overview of service techniques. The pictures and illustrations that are used in this chapter are for demonstration purposes to clarify the description of how to service ice makers. They in no way reflect on a particular brand's reliability.

PRINCIPLES OF OPERATION FOR TYPE-1 AND TYPE-2 AUTOMATIC ICE MAKERS

The freezer temperature determines the efficiency of the automatic ice maker. The colder the temperature in the freezer, the faster ice will freeze in the mold or tray. In order for the ice maker to harvest the ice cubes, the temperature in the freezer should be colder than 12 degrees Fahrenheit.

The water supply (Fig. 14-1) is metered into the ice maker mold or tray by a water valve, usually located in the compressor compartment. The water freezes in the automatic ice maker mold or tray, as it would in a manual ice cube tray.

Type-1 ice maker

Type-1 ice makers, as illustrated in Fig. 14-2, have a thermostat that is mounted on the ice maker mold which will sense the temperature of the ice. When the water and ice has cooled down to 17 ± 3 degrees Fahrenheit, the thermostat will activate the ice maker motor and heater circuit. The motor will rotate the ejector blade, allowing it to rest on the ice cubes and, at the same time, exerting pressure. The ice maker

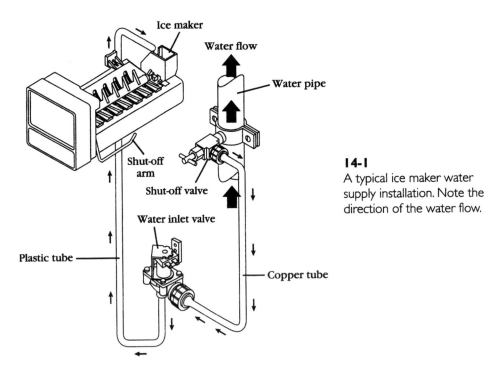

14-1
A typical ice maker water supply installation. Note the direction of the water flow.

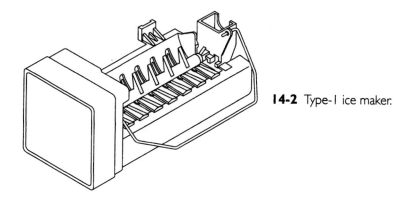

14-2 Type-1 ice maker.

mold heater will warm up the ice cubes just enough to release the ice cubes from the mold. This procedure will take approximately three to five minutes to release the ice cubes from the mold.

When the ice cubes become free from the ice maker mold, the ejector blade will continue to rotate, scooping the ice cubes out of the ice maker mold and depositing them in the ice bucket. The shut-off arm moves up and down as the ice maker cycles. At the end of the cycle, the arm will lay on top of the ice cubes. As the cubes fill the ice bucket, the shut-off arm raises to a designated point, and turns off the ice maker; halting ice production.

As the ice level in the bucket falls, from use, the ice maker cycle will resume. Then, the water valve opens again, allowing the water to enter the ice maker mold to be frozen into ice cubes; thus, beginning the cycle all over again. The fill time for the type-1 ice maker is approximately seven seconds. This type of ice maker can produce up to eight pounds of ice within a 24-hour period, providing the conditions are ideal. Some customers will have ice production between 3 to 5 pounds per day, depending on usage and door openings.

Type-2 ice maker

The type-2 ice maker (Fig. 14-3) has the components in the ice maker head necessary for the ice-making operation. The temperature in the freezer compartment must be below 15 degrees Fahrenheit before the ice maker will begin to operate. Inside the ice maker head assembly is a thermostat that senses the temperature in the freezer compartment. When the thermostat is satisfied, it will energize the timer motor. When the motor is energized, it will begin to turn a timing gear, and the shut-off arm will also begin to move. The shut-off arm will descend into the ice bucket to determine how much ice is in the bucket. If the bucket is full, the shut-off arm will rest against the ice cubes and the ice maker will not cycle. If the shut-off arm continues to return to its normal position, the ice bucket is not full, and the cycle will continue.

If the cycle continues, the tray will begin to rotate. When the tray has rotated about 140 degrees, one corner of the tray engages the tray stop, which prevents that part of the tray from rotating further (Fig. 14-4). The rest of the tray will continue to rotate and twist the tray to about 40 degrees. At this point, the twisting of the tray will loosen the ice cubes. As the shaft continues to turn, the tray stop will begin to retract. As this happens, the tray will rapidly release the ice cubes into the ice

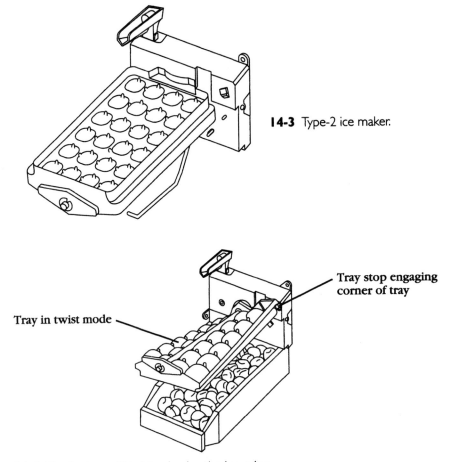

14-3 Type-2 ice maker.

Tray stop engaging corner of tray

Tray in twist mode

14-4 The ice tray will twist, releasing the ice cubes.

bucket. After the ice has been released, the tray will continue to rotate until it has completed its 360-degree rotation. Near the end of the tray rotation, the water valve fill switch is energized, allowing the water to enter and fill the tray for the next cycle (Fig. 14-5). The fill time for type-2 ice maker is approximately 12 seconds.

Testing type-1 ice maker

The type-1 ice maker is designed to allow all of the components to be tested without removing the ice maker, and without moving the refrigerator away from the wall to test the water valve.

The ice maker in Fig. 14-6A is the old-style ice maker, and it is not in production anymore. Parts for this old style ice maker are still available. Figure 14-6B is the new-style modular ice maker. The components that make up the new-style modular ice maker are illustrated in Fig. 14-7.

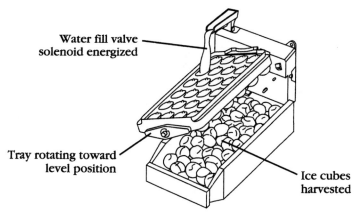

Water fill valve
solenoid energized

Tray rotating toward
level position

Ice cubes
harvested

14-5 The ice cube tray must return to a level position after the filling cycle is complete.

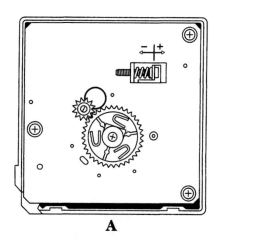

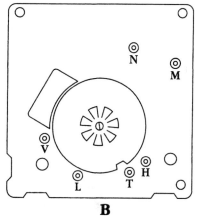

A

B

14-6 A. Old-style type-1 ice maker. B. New-style type-1 ice maker.

The test holes that are on the ice maker head module (Fig. 14-6B) are identified as "N," "M," "H," "T," "L," "V." These are the test points for testing this type of ice maker. The letters indicate the following:

N = neutral side of line voltage
M = ice maker motor connection
H = mold heater connection
L = L1 side of line voltage
V = ice maker water valve connection. This test requires the electricity to be turned on.

Caution should be taken when working with live wires. Avoid getting shocked. Stay clear of live wires. Only handle the meter probes by the insulated handles.

To test this ice maker, the unit must be installed and plugged into the freezer ice maker receptacle; the shut-off arm placed in the down position; and the tempera-

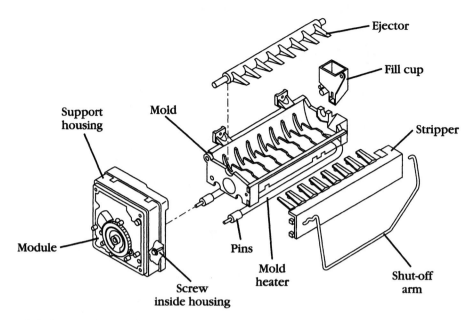

Ejector

Fill cup

Support housing

Mold

Stripper

Module

Pins

Screw inside housing

Mold heater

Shut-off arm

14-7 Exploded view of new-style type-1 ice maker.

ture in the freezer should be colder than 12 degrees Fahrenheit. Set your multimeter for voltage and place one probe in test point "L" and the other probe in test point "N." Be sure that the test probes go into the test points about ½ inch. You should have a reading of 120 volts, which indicates line voltage to the ice maker. Next, place the meter probes in test points "T" and "H." This will verify if the bimetal thermostat is open or closed. If the thermostat is open, you will read 120 volts. If it is closed, you will read no line voltage. At this point, you are going to use an insulated jumper wire (Fig. 14-8) to short the test points "T" and "H."[1] This procedure will run the motor. If the motor doesn't run, replace the ice maker module assembly. If the ice maker motor runs, replace the bimetal thermostat. If you leave the jumper wire in for a half of a revolution, you can feel the mold heater heating up. This means that the mold heater is working. Now, remove the jumper wire, and the water valve will be energized in the last half of the revolution. The ejector blade in Fig. 14-9 is rotating clockwise. This illustration shows what is about to happen as the cycle begins (from the stop position) and rotates 360 degrees back again.

Another way to test the water valve, without cycling the ice maker, is to place one end of the jumper wire in test point "N" and the other end in test point "V." Water will immediately enter the ice maker mold, so be ready to disconnect the jumper wire. If no water enters the mold, then check the water supply, the water valve, and the connecting tubing.

---

[1] The jumper wire should be made of 14 gauge wire, approximately 8 to 10 inches long, with ⅜ inch of the insulation removed from both ends. Do not short any contacts other than those specified in this test procedure. Why? Because you can damage the ice maker, or injure yourself, or both.

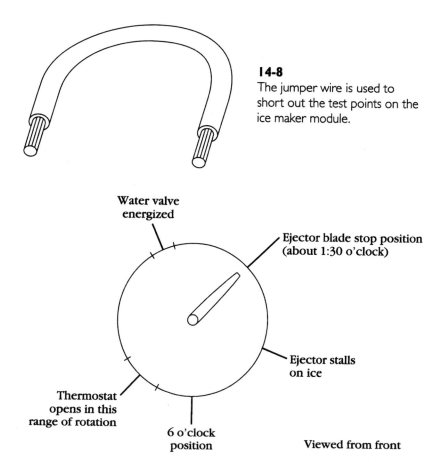

14-8
The jumper wire is used to short out the test points on the ice maker module.

14-9 Indicating the ejector blade position when ice maker is cycling.

Testing type-2 ice maker

In testing the operation of the type-2 ice maker head, you will have to initiate the harvest cycle. Do not remove the ice maker for this test. The shut-off arm will not operate during the manual harvest. When performing the following procedures refer to Fig. 14-10. To cycle the ice maker do the following:

1. Place the shut-off arm in the on position.
2. Push the switch actuator tab down, and hold it. This will activate the ice maker motor.
3. Push the tray lock tab toward the tray shaft. This will unlock the tray.
4. Twist the tray clockwise. This will start the cycle.
5. As the tray turns past 30 degrees, you can release the switch actuator tab.
6. After the harvest cycle is complete, empty the water from the ice tray. This will prevent the next occurring automatic harvest from dumping water into the ice bucket. When you perform a manual harvest, you interrupt the timing sequence only. A manual harvest will not reset the timing mechanism.

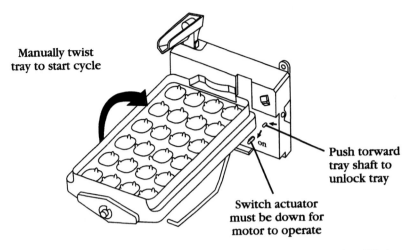

Manually twist tray to start cycle

Push torward tray shaft to unlock tray

Switch actuator must be down for motor to operate

14-10 Manually twist the ice tray clockwise. When the tray reaches 30 degrees, then release the switch actuator.

If you are unable to start the harvest cycle, then check the motor shaft to see if it turns while you depress the switch actuator (Fig. 14-11). Also, check the temperature in the freezer. If the temperature is above 15 degrees Fahrenheit, the ice maker will not operate. If the motor shaft still does not rotate, then remove the ice maker, and test the unit on the work bench. Use a test cord to test the ice maker motor (Fig. 14-12).[2]

Water/temperature problems

The water quality can cause the ice maker to malfunction or produce poor-quality ice cubes. If the minerals in the water, or sand particles, become lodged in the water

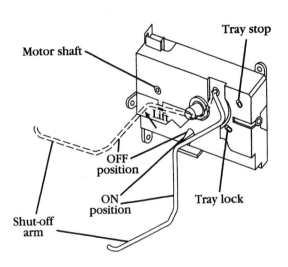

Motor shaft

Tray stop

Lift

OFF position

ON position

Tray lock

Shut-off arm

14-11 To shut off the ice maker, lift the shut-off arm and rest it on the tab, as indicated by the broken lines.

---

[2] You can purchase this ice maker test cord from your parts distributor.

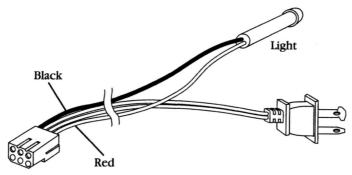

14-12 A test cord to test a type-2 ice maker.

valve fill screen, it can restrict or stop the water from entering the ice maker mold. If sand particles bypass the fill screen in the water valve, it can possibly keep the water valve from closing completely, thus causing the water to enter the ice maker mold continuously. If this condition happens, the ice maker will produce small cubes or even flood the ice maker compartment and freezer. Another indication of a defective water valve is that the ice maker fill tube will be completely frozen shut, possibly causing water to flow onto the floor every time the ice maker cycles.

The minerals in the water can also cause lime deposits to build up on the ice maker mold. If this condition happens, the ice cubes will not be released easily from the mold.

Ice production can be slowed if the temperature in the freezer compartment is above normal. To correct this problem, adjust the temperature controls to a colder position. The more the doors are opened, the colder the temperature setting must be.

STEP-BY-STEP TROUBLESHOOTING TYPE-1 ICE MAKERS

When servicing an appliance, don't overlook the simple things that might be causing the problem. Step-by-step troubleshooting, by symptom diagnosis, is based on diagnosing malfunctions with possible causes arranged in categories relating to the operation of the automatic ice maker. This section is intended only to serve as a checklist, to aid you in diagnosing a problem. Look at the symptom that best describes the problem you are experiencing with the automatic ice maker, then correct the problem.

Ice maker producing too much ice

1. Check if the shut-off arm is not connected to the actuator in the ice maker head.
2. Check the shut-off arm. Is it bent out of its original shape?
3. Check the shut-off linkage in the ice maker head. Is it broken?

Ice maker will not make ice, or low ice production

1. Check freezer temperature.
2. Is the shut-off arm in the off position?
3. Check motor operation. Did it stall? Are the gears stripped?

4. Test the voltage at the ice maker. Is the ice maker plugged into the receptacle?
5. Test thermostat for continuity. Bypass thermostat. Will the ice maker run?
6. Check the shut-off linkage in the ice maker head. Is it broken?
7. Is the supply water turned on? Does water enter the ice maker mold when it cycles?
8. Check the ice maker fill tube. Is it frozen?
9. Check for an ice jam.
10. Check for defective wiring.

Ice cubes too small

1. Check the ice maker. Is it level?
2. Check the fill tube. Is it frozen?
3. Check water supply.
4. Check ice maker water valve.
5. Check the water pressure. Is it between 20 to 120 PSI?
6. Check for a self piercing saddle valve. Mineral deposits will restrict the opening.
7. Cycle the ice maker and catch the fill water in a glass. Measure the amount of water. Are there at least 140 cc of water in the glass?
8. Test for opens in the mold heater.
9. Check the ice maker thermostat. There might be insufficient thermal bond between the thermostat and the ice maker mold.

Ice maker producing hollow ice cubes

1. Cycle the ice maker, and catch the fill water in a glass. Measure the amount of water. Are there at least 140 cc of water in the glass?
2. Check for improper air flow in the freezer compartment. Direct the air flow away from the ice maker thermostat.
3. Check the ice maker thermostat.
4. Check the temperature in the freezer compartment.

Ice maker flooding the freezer compartment or ice bucket

1. Check the thermostat.
2. Check for an ice jam, when the ice maker is in the fill position.
3. Check for a leaky water valve.
4. Check for proper ice maker water fill. Too much water will spill over the mold, causing the ice in the bucket to freeze together.
5. Check the ejector blade position. If the blade is in the 12 o'clock position, the ice maker motor has stalled.
6. Check ice maker module for contamination and/or burned switch contacts. Check the linkage for proper operation.
7. Is the refrigerator level? Is the ice maker level?
8. Check water pressure. Ice makers fill according to time, not volume. Water pressure should be between 20 to 120 PSI.
9. Is the fill tube located in the fill cup?

STEP-BY-STEP TROUBLESHOOTING TYPE-2 ICE MAKER BY SYMPTOM DIAGNOSIS

When servicing an appliance, don't overlook the simple things that might be causing the problem. Step-by-step troubleshooting, by symptom diagnosis, is based upon diagnosing malfunctions with possible causes arranged into categories relating to the operation of the automatic ice maker. This section is intended only to serve as a checklist, to aid you in diagnosing a problem. Look at the symptom that best describes the problem you are experiencing with the automatic ice maker, then proceed to correct the problem.

Ice maker will not run

1. Test freezer temperature.
2. Check the shut-off arm. Is it in the off position?
3. Run ice maker test. Does the ice maker motor run?
4. Check for defective wiring.
5. Is the ice maker plugged into the receptacle?
6. Test for voltage at the ice maker receptacle.

Ice cubes are stuck together

1. Test temperature of freezer.
2. Check for proper fill.
3. Check ice tray for mineral deposits. Mineral deposits will cause the ice to stick to the tray, which can cause the ice cubes to stick together on the next cycle.

Ice maker spills water from the tray

1. Check the ice tray. When the harvest cycle is completed, the ice cube tray should return to its starting position, and the tray should be level.
2. Check the inlet water fill tube. Be sure that the fill tube and fill trough fit together properly.
3. Check for a leaking water inlet valve.

Water will not enter ice tray

1. Check for proper water supply to ice maker.
2. Check water valve strainer for restrictions.
3. Check the water valve.
4. Check for proper water pressure.
5. Check wiring circuit.

Ice cubes too small, or some of the ice cube compartments are empty

1. Check the ice maker. Is it level?
2. Check the fill tube. Is it frozen?
3. Check water supply.

4. Check ice maker water valve.
5. Check the water pressure. Is it between 20 to 120 PSI?
6. Check for a self piercing saddle valve. Mineral deposits will restrict the opening.
7. Cycle the ice maker and catch the fill water in a glass. Measure the amount of water. Are there at least 200 cc of water in the glass? The fill time will be between 12 and 13 seconds.

DIAGNOSTIC CHART

The diagnostic chart (Fig. 14-13) will help you to pinpoint the likely cause of the problem.

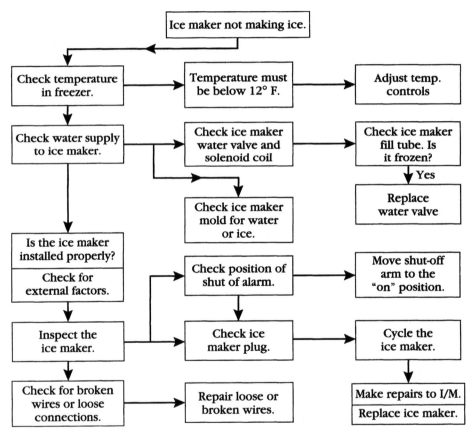

14-13 The diagnostic flow chart: ice maker not making ice.

REPAIR PROCEDURES

Each repair procedure is a complete inspection and repair process for a single ice cube maker component, containing the information you need to test and replace components.

Type-1 ice maker, module

The typical complaints associated with ice maker module failure are: 1. Ice maker not making any ice. 2. Ice maker stalls in middle of cycling.

1. *Verify the complaint* Verify the complaint by checking the ice maker and the temperature in the freezer.
2. *Check for external factors* You must check for external factors not associated with the ice maker. For example: Is the ice maker installed properly?
3. *Disconnect the electricity* Before working on the ice maker, disconnect the electricity to the refrigerator/freezer. This can be done by pulling the plug from the receptacle. Or disconnect the electricity at the fuse panel or at the circuit breaker panel. Turn off the electricity.
4. *Gain access to the ice maker* Open the freezer door to access the ice maker. Remove the screws that secure the ice maker to the wall of the freezer (Fig. 14-14). Next, unplug the ice maker from the receptacle and remove the ice maker (Fig. 14-15A). Now, depress the retaining tab, and pull the wiring harness from the ice maker head (Fig. 14-15B).

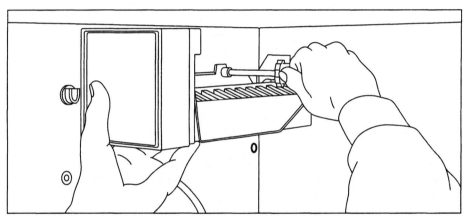

14-14 Support the ice maker when removing the screws.

5. *Disassemble and remove the ice maker module* After testing the ice maker, and it has been determined that you must replace the ice maker module, it is time to disassemble it. Insert a flat blade screwdriver between the shut-off arm and the white elongated hole on the ice maker head (Fig. 14-16), remove the shut-off arm by prying the arm out of the ice maker head. Pry off the ice maker cover (Fig. 14-17) with a coin or a screwdriver. This will expose the ice maker module. To remove the ice maker module, remove the three screws (Fig. 14-18) and pull the module out of the ice maker head. Inspect the module linkage and switch contacts (Fig. 14-19).
6. *Install a new ice maker module* To install the new ice maker module, just reverse the order of disassembly, and reassemble. To reinstall the shut-off arm on the ice maker (Fig. 14-16), you must first insert the straight end in the round hole in the fill cup. Be sure the flat side on the arm goes through

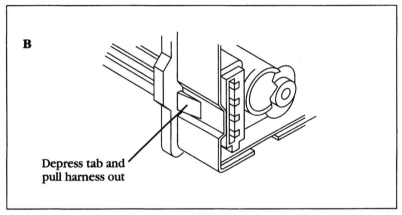

14-15 Pull ice maker plug out of receptacle (A). Remove the wire harness by depressing the tab, then pull it (B).

the fill cup. This will prevent the shut-off arm from coming out of the fill cup hole. Next, insert the other end of the shut-off arm into the white elongated hole in the ice maker housing. *Do not* install the arm into any one of the *round* holes in the ice maker housing. Push on the arm so that the arm will be completely in place, and even with the ice maker surface. Reinstall the ice maker in the freezer and test it.

Type-I ice maker, thermostat

The typical complaint associated with ice maker thermostat failure are: 1. Ice maker not making any ice cubes. 2. Ice maker producing hollow ice cubes.

    1. *Verify the complaint* Verify the complaint by checking the ice maker and the temperature in the freezer.

    2. *Check for external factors* You must check for external factors not associated with the ice maker. Is the ice maker installed properly?

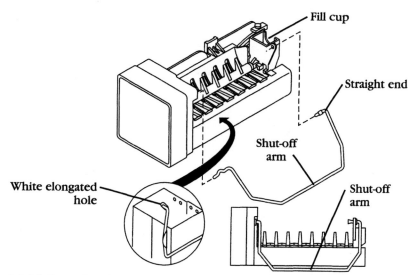

14-16 Removing the shut-off arm by prying out of slot. When installing the shut-off arm, be sure that the end is pressed in all the way.

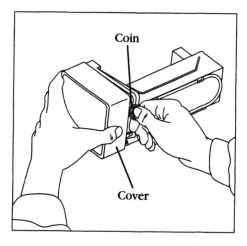

14-17
Gain access to the ice maker module.

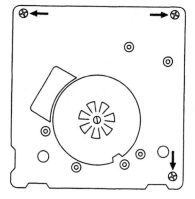

14-18
Removing the three screws to remove the module.

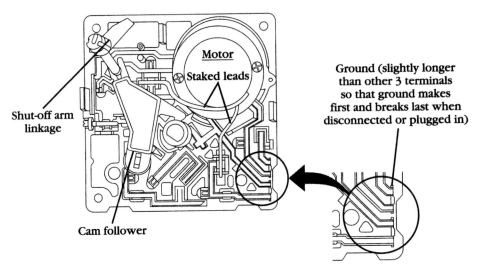

Motor
Staked leads

Ground (slightly longer
than other 3 terminals
so that ground makes
first and breaks last when
disconnected or plugged in)

Shut-off arm
linkage

Cam follower

14-19 Inspecting the shut-off arm linkage, cam follower, and switch contacts. Note: the ground terminal is slightly longer than the other three terminals. The ground connection is made first, or breaks last, when the wiring harness is removed or plugged in.

3. *Disconnect the electricity* Before working on the ice maker, disconnect the electricity to the refrigerator/freezer. This can be done by pulling the plug from the receptacle. Or disconnect the electricity at the fuse panel or at the circuit breaker panel. Turn off the electricity.

4. *Gain access to the ice maker thermostat* Open the freezer door to access the ice maker. Remove the screws that secure the ice maker to the wall of the freezer (Fig. 14-14). Next, unplug the ice maker from the receptacle, and remove the ice maker (Fig. 14-15A). Now, press the retaining tab and pull the wiring harness from the ice maker head (Fig. 14-15B). Insert a flat blade screwdriver between the shut-off arm and the white elongated hole on the ice maker head (Fig. 14-16), remove the shut-off arm by prying the arm out of the ice maker head. Pry off the ice maker cover (Fig. 14-17) with a coin or a screwdriver. This will expose the ice maker module. To remove the ice maker module, remove the three screws (Fig. 14-18) and pull the module out of the ice maker head. Inspect the linkage, the cam follower, and the switch contacts. Next, remove the two screws that secure the ice maker head to the mold (Fig. 14-20). Separate the ice maker head from the mold assembly and you will see the thermostat on the mold side of the ice maker head.

5. *Remove the thermostat* To remove the thermostat, use needle nose pliers to pull out the retaining clips (Fig. 14-21). Remove the thermostat.

6. *Install a new thermostat* To install the new thermostat, just reverse the order of disassembly, and reassemble the ice maker. Be sure that you properly index the pins on the thermostat. Before you assemble the ice maker head to the mold assembly, you must apply new thermal bonding material to the thermostat. This will allow the thermostat to make better contact with the ice maker mold.

Reinstall the ice maker in the freezer, and test.

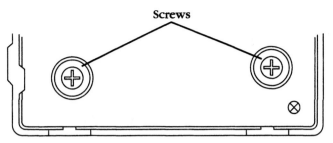

14-20 Removing the two screws.

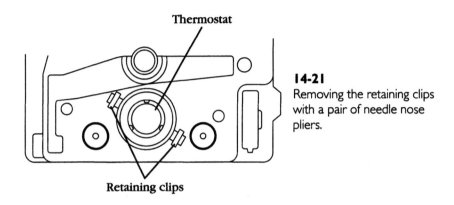

Thermostat

14-21
Removing the retaining clips
with a pair of needle nose
pliers.

Retaining clips

Type-1 ice maker, mold, and heater assembly

The typical complaints associated with ice maker mold and heater assembly failure
are: 1. Ice maker not making any ice. 2. Ice will not come out of the mold. 3. Black
pieces of mold coating material in the ice cubes.

1. *Verify the complaint* Verify the complaint by checking the ice maker and
   the temperature in the freezer.
2. *Check for external factors* You must check for external factors not
   associated with the ice maker. Is the ice maker installed properly?
3. *Disconnect the electricity* Before working on the ice maker, disconnect the
   electricity to the refrigerator/freezer. This can be done by pulling the plug
   from the receptacle. Or disconnect the electricity is at the fuse panel or
   circuit breaker panel. Turn off the electricity.
4. *Gain access to the ice maker mold assembly* Open the freezer door to access
   the ice maker. Remove the screws that secure the ice maker to the wall of
   the freezer (Fig. 14-14). Next, unplug the ice maker from the receptacle, and
   remove the ice maker (Fig. 14-15A). Now, press the retaining tab and pull
   the wiring harness from the ice maker head (Fig. 14-15B).
      Insert a flat blade screwdriver between the shut-off arm and the white
   elongated hole on the ice maker head (Fig. 14-16), remove the shut-off
   arm by prying the arm out of the ice maker head. Pry off the ice maker
   cover (Fig. 14-17) with a coin or screwdriver. Use a phillips screwdriver,

and insert it into the access ports in the module (Fig. 14-22) to loosen the screws. Separate the ice maker head assembly from the mold assembly. Also, remove the ejector, the fill cup, the stripper, and the shut-off arm (Fig. 14-7). Do not remove the heater. The heater and mold come as a complete assembly.

5. *Install a new ice maker mold assembly* To install new ice maker mold assembly, just reverse the order of disassembly, and reassemble. When assembling the ice maker, apply a thin film of silicone grease to the end of the ejector that goes into the fill cup. This will prevent the ejector from freezing to the fill cup. Also, apply a thin film of silicone grease to the other end of the ejector. Before installing the stripper, apply a heavy film of silicone grease to the top surface of the mold that is covered by the stripper. This will prevent the water from wicking over the ice maker mold every time it cycles. Before you assemble the ice maker head to the mold assembly, you must apply new thermal bonding material to the thermostat. This will allow the thermostat to make better contact with the ice maker mold. Reinstall the ice maker in the freezer and test it.

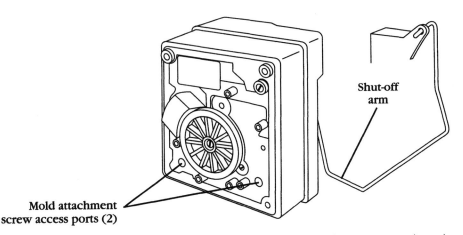

14-22 The entire ice maker head can be removed by loosening the two screws through the access ports.

Water valve

The typical complaints associated with water valve failure are: 1. Ice maker not making any ice. 2. Ice maker fill tube frozen. 3. Ice maker mold fills with very little water.

1. *Verify the complaint* Verify the complaint by checking the ice maker and the temperature in the freezer.

2. *Check for external factors* You must check for external factors not associated with the ice maker. Is the ice maker installed properly? Check the water supply to the water valve.

3. *Disconnect the electricity* Before working on the water valve, disconnect the electricity to the refrigerator/freezer. This can be done by pulling the

plug from the receptacle. Or disconnect the electricity at the fuse panel or circuit breaker panel. Turn off the electricity.

4. *Gain access to the water valve* To access the water valve, pull the refrigerator/freezer away from the wall. Remove the back access panel. Shut off the water supply to the ice maker.

5. *Remove the water valve* Remove the water inlet tube from the water valve. Then, remove the screws from the water valve bracket that secure the valve to the cabinet (Fig. 14-23). Next, remove the ice maker fill line. Finally, disconnect the wiring harness from the solenoid coil of the valve.

6. *Test the water valve* Using your ohmmeter, set the scale on R × 10, and place the probes on the solenoid coil terminals (Fig. 14-24). The meter should show between 200 and 500 ohms resistance. If not, replace the water valve. If you determine that the water valve is good, but there is little water flow through the valve, inspect the inlet screen. If this screen is filled with debris, it must be cleaned out. To accomplish this, use a small flat blade screwdriver and pry out the screen (Fig. 14-25). Then, wash out the screen, being sure all of the debris is removed. After cleaning out the debris, reinstall the screen, and test the water valve.

7. *Install a new water valve* To install the new water valve, just reverse the disassembly procedure, and reassemble. Also, reconnect the ground wire from the water valve bracket to the refrigerator/freezer cabinet. Secure the water supply line to the cabinet. This will prevent the supply line from leaking when the refrigerator/freezer is pushed back against the wall. Check for water leaks before you push the refrigerator/freezer back against the wall. Also inspect the fill tubing for any cracks, etc. (Fig. 14-26).

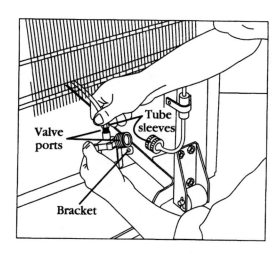

14-23
Disconnect the water inlet line and the outlet line. Disconnect wiring harness from the solenoid coil.

Type-2 ice maker head and tray

The typical complaints associated with type-2 ice maker head failure are: 1. Ice maker not making any ice. 2. Ice will not come out of the mold.

1. *Verify the complaint* Verify the complaint by checking the ice maker and the temperature in the freezer.

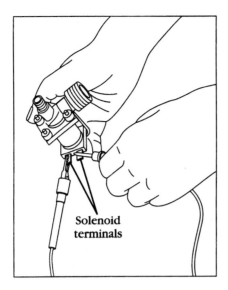

14-24
Testing the solenoid coil for resistance.

Solenoid
terminals

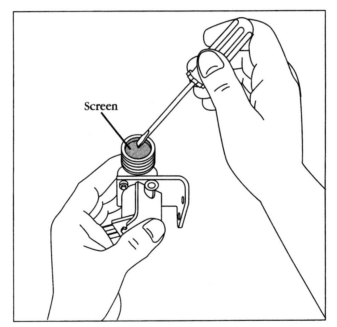

14-25 Removing the inlet screen. Clean it out with warm water and soap.

2. *Check for external factors* You must check for external factors not associated with the ice maker. Is the ice maker installed properly?

3. *Disconnect the electricity* Before working on the ice maker, disconnect the electricity to the refrigerator/freezer. This can be done by pulling the plug

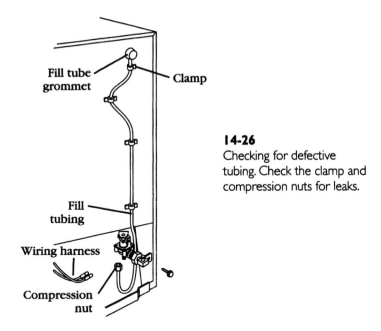

14-26
Checking for defective tubing. Check the clamp and compression nuts for leaks.

from the receptacle. Another way to disconnect the electricity is at the fuse panel or circuit breaker panel. Turn off the electricity.

4. *Gain access to the ice maker* Open the freezer door to access the ice maker. Remove the screws that secure the ice maker to the wall of the freezer. Next, unplug the ice maker from the receptacle and remove the ice maker.

5. *Disassemble and remove the ice maker head assembly* After testing the ice maker, and it has been determined that you must replace the ice maker head assembly, it's time to disassemble it. First, remove the retaining clip from the end of the tray, and then slide the tray off the ice maker drive shaft (Fig. 14-27). Next, remove the fill cup and screw. On the back of the ice maker head is a clamp that holds the tray drive shaft in place; remove it. Pull the shaft out of the ice maker head. Then, place a flat blade screwdriver behind the shut-off arm and pry it loose (Fig. 14-28).

6. *Install a new ice maker head* To install the new ice maker head, just reverse the order of disassembly, and reassemble. When replacing a defective ice maker head, only replace it with an exact replacement having the same amount of cycle operation time. Reinstall the ice maker in the freezer and test.

Type-2 ice maker tray

The typical complaints associated with type-2 ice maker tray failure are: 1. Ice maker not making any ice. 2. Ice will not come out of the mold. 3. Water in the tray freezes into a solid block of ice.

1. *Verify the complaint* Verify the complaint by checking the ice maker and the temperature in the freezer.

2. *Check for external factors* You must check for external factors not associated with the ice maker. Is the ice maker installed properly?

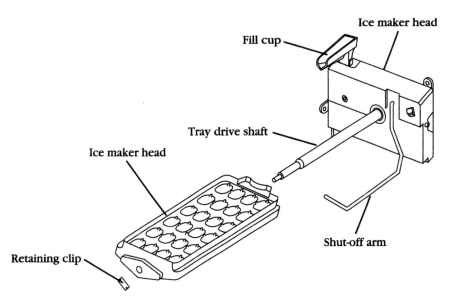

Ice maker head

Fill cup

Tray drive shaft

Ice maker head

Shut-off arm

Retaining clip

14-27 Always reinstall the ice cube tray in the same position as when it was removed. If the ice tray is installed upside down, water will flood the ice bucket every time the ice maker cycles.

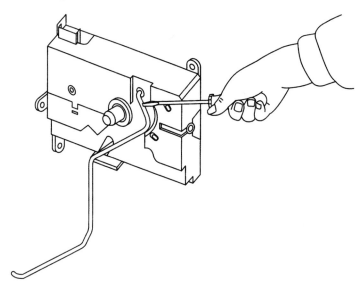

14-28 Gently pry off the shut-off arm.

3. *Remove the ice maker tray* To remove the ice maker tray, remove the retaining clip from the end of the tray, and then slide the tray off the ice maker drive shaft (Fig. 14-27).
4. *Install a new ice tray* To install the new ice tray, just reverse the disassembly procedure, and reassemble. Always reinstall the tray in the

same position, as that from which it was removed, for proper operation. Test the ice maker. Let it run through a complete cycle and harvest. Then, inspect the ice maker tray.

SELF-CONTAINED ICE CUBE MAKER

This type of ice cube maker is a freestanding, self-contained refrigeration appliance which produces a slab of ice that is cut into ice cubes (Fig. 14-29). The production of ice cubes is all done automatically, and the entire mechanism is stored within the ice cube maker cabinet. The self contained ice cube maker can also be installed under the counter. The thickness of the ice cubes can be adjusted by the thickness control, located on the control panel. This ice cube maker can produce up to 50 pounds of ice cubes per 24 hours. The amount of ice cubes will vary, depending on where it is installed, the room temperature, and the supply water temperature.

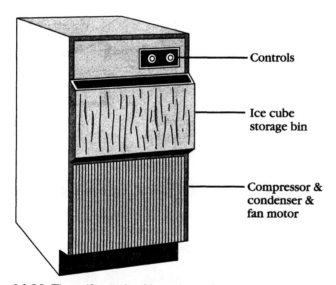

14-29 The self-contained ice cube maker.

Principles of operation

In the freeze cycle, water flows constantly, and it is recirculated over the evaporator freeze plate until a slab of ice is formed (Fig. 14-30). When the ice slab reaches a predetermined thickness, the evaporator freeze plate temperature is sensed by the thermostat, and the freeze cycle is terminated. At that point in time, the defrost cycle will begin to release the ice slab from the evaporator freeze plate. The ice slab will slide down onto the cutter grid, which cuts the ice slab into ice cubes. At the same time, the ice maker has automatically switched back into the freeze cycle.

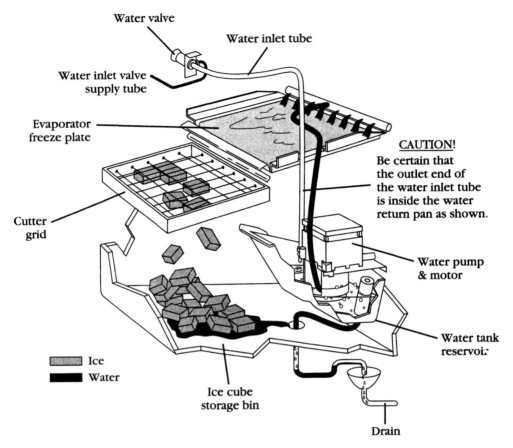

Water valve

Water inlet tube

Water inlet valve
supply tube

Evaporator
freeze plate

CAUTION!
Be certain that
the outlet end of
the water inlet tube
is inside the water
return pan as shown.

Cutter
grid

Water pump
& motor

Water tank
reservoir

Ice

Water

Ice cube
storage bin

Drain

14-30 A pictorial view of the water system.

When the defrost cycle begins, the remaining water in the water tank reservoir is discarded out through the overflow tube. At that point, fresh water will enter through the water valve, and go into the water tank reservoir.

The refrigeration cycle (Fig. 14-31) in this type of ice maker is similar to the conventional refrigerator. The compressor pumps the refrigerant into the condenser coil, which is cooled by a fan and motor. The refrigerant leaves the condenser coil as a high-pressure liquid, passes through the dryer and enters into the capillary tube. The refrigerant is next metered through the capillary tube, and it then enters the evaporator freeze plate. The refrigerant gas will then leave the evaporator freeze plate and return to the compressor.

When the ice maker goes into the defrost cycle, it energizes the hot gas solenoid, which reverses the refrigeration cycle, during which the condenser fan motor and water pump will stop. The hot gas passes through the evaporator freeze plate, heating it up enough to release the ice slab. The thermostat will sense the temperature of the evaporator freeze plate again, and it will activate the freeze cycle. The hot gas solenoid valve will then close, the water valve will close, the condenser fan

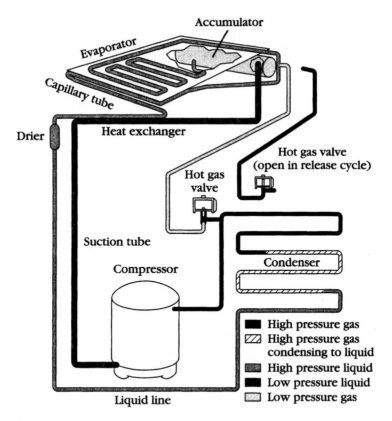

14-31 A pictorial view of the refrigeration system.

motor will start, the water pump will start, and the freeze cycle will begin to manufacturer a new slab of ice.

STEP-BY-STEP TROUBLESHOOTING BY SYMPTOM DIAGNOSIS

When servicing the self-contained ice cube maker, don't overlook the simple things that might be causing the problem. Step-by-step troubleshooting, by symptom diagnosis, is based on diagnosing malfunctions, with possible causes arranged into categories relating to the operation of the automatic ice maker. This section is intended only as a checklist, to aid you in diagnosing a problem. Look at the symptom that best describes the problem you are experiencing with the automatic ice maker, then correct the problem.

Compressor will not run; no ice in storage bin

1. Is the ice cube maker located in an area where the temperature is below 55 degrees Fahrenheit ?
2. Test for proper voltage supply.

3. Check for loose or broken wires.
4. Test the compressor, relay and overload protector.
5. Check the controls for the proper setting.
6. Test bin thermostat for continuity. If contacts are open, replace the thermostat.
7. Test the compressor.

Compressor runs; no ice in storage bin

1. Check water supply.
2. Check water valve.
3. Check evaporator thermostat.
4. Check the hot gas solenoid. It might be stuck "open."
5. Check for sealed-system problems.
6. Check for excessive use of ice cubes.
7. Test cutter grid.
8. Check wiring against wiring diagram.
9. Is the water inlet tube from the water valve inserted in the return trough?
10. Check condenser fan motor.

Ice storage bin full of ice; compressor runs continuously

1. Check the calibration on the bin thermostat.
2. Test the bin thermostat for continuity. Are the contacts stuck shut?
3. Check wiring against the wiring diagram.

Low ice production

1. Is the ice cube maker located in an area where the temperature is below 55 degrees Fahrenheit?
2. Inspect the storage bin. Is water falling on the ice cubes?
3. Check the calibration on the bin thermostat.
4. Check the thickness control. Ice cubes produced should be between ½ to ⅝ inch thick.
5. Check the hot gas solenoid. It might be partially stuck open.
6. Check for sealed-system problems.
7. Check water supply in reservoir. There might not be enough water circulating over the evaporator freeze plate.

Excessive water dripping on the ice cubes

1. Is the water tank overflowing? Check for blocked overflow tube.
2. Is the water trough installed properly?
3. Is the water inlet tube from the water valve inserted in the return trough ?
4. Check cutter grid for ice jam.
5. Check the water deflector position.
6. Check water valve for leaks.

Ice cubes too thin

1. Check thickness control setting.
2. Check if there is enough water being circulated over the evaporator freeze plate.
3. Check for restrictions in the water system.
4. Check the water pump, motor, and the distributor tube.
5. Check the thermostat calibration.

Ice cubes too thick

1. Check thickness control setting.
2. Check the thermostat calibration.

The condenser fan will not run during the freeze cycle

1. Check the fan blades for binding on the shroud.
2. Test the condenser motor for continuity.
3. Check for open circuits against wiring diagram.
4. Check for a defective evaporator thermostat.

Water pump will not run

1. Check the pump for binding in the housing.
2. Check for open circuits against wiring diagram.
3. Test the pump motor for continuity.
4. Check for a defective evaporator thermostat.

Water tank is empty

1. Water will not enter tank until the first defrost cycle is initiated.
2. Check for open circuit to the water valve solenoid.
3. Check water line for complete restriction.
4. Check for a defective evaporator thermostat.
5. Check the water valve, it might be stuck shut.
6. Test the water valve solenoid for continuity.
7. Check the water inlet tube from the water valve. It might not be directing the water into the tank.
8. Check for a clogged water inlet screen in the water valve.

Treating the water

In the freeze cycle, as the water passes over the evaporator freeze plate, the impurities in the water are rejected and only the pure water will stick to the plate. The more dissolved solids that are present in the water, the longer the freezing cycle. Bicarbonates, which are found in the water, are the most troublesome of all of the impurities. These impurities can cause:

- Scaling on the evaporator freeze plate
- Clogging of the water distributor head

• Water valve, and many other parts in the water system, to clog up

If the impurities become too concentrated in the water system, they can cause:

• Cloudy cubes
• Mushy ice

All of the water system parts that come in direct contact with the water might become corroded if the water supply is high in acidity.

The water might have to be treated in order to overcome the problems with the mineral content. The most economical way to treat the water supply is with a polyphosphate feeder. This feeder is installed in the water inlet supply to prevent scale build-up. This will require less frequent cleaning of the ice maker. To install one of these feeders, follow the manufacturer's recommendations in order to treat the water satisfactorily.

The manufacturer of this type of ice cube maker recommends that the ice maker be cleaned occasionally to help combat lime and mineral deposit build-up.

CLEANING INSTRUCTIONS FOR THE ICE CUBE MAKER

To clean the water system parts and the evaporator freeze plate, turn off the ice maker with the cycle switch. Open the bin door and remove the cutter grid by removing the two thumb screws, unplug the cutter grid, and remove it from the storage bin (Fig. 14-32). A drain plug is located under the water tank. Remove it to drain the water out of the tank (Fig. 14-33). After all of the water has been removed, reinstall the plug. Pour ½ gallon of hot water into the tank and set the switch to clean. The hot water will circulate through the water pump assembly and over the evaporator freeze plate, including all the water systems components. Let the water circulate for five minutes, then drain the water out of the tank. Replace the plug. Mix ice machine cleaner with ½ gallon of hot water, and pour it into the water tank. If you use a recognized ice machine cleaner, follow the instructions on the label for best results. If you would rather prepare your own solution, add 6 ounces of citric acid and phosphoric acid to ½ gallon of hot water, and pour into the water tank.

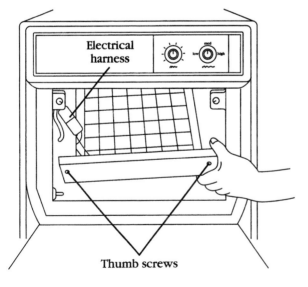

14-32
Removing the thumbscrews, disconnecting the electrical harness, and pulling the cutter grid out of the bin.

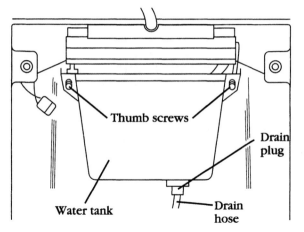

14-33
The water tank is located in the storage bin.

Thumb screws

Drain plug

Water tank

Drain hose

Turn the switch to clean and circulate this solution for 20 minutes or longer, then drain the water. Follow with two clean water rinses that circulate for five minutes, and then drain the water.

Remove the splash guard, the water dispenser tube, and the plastic water pump tank; then place them in a solution of mild laundry bleach for five minutes, and then rinse. Use one ounce of bleach to one gallon of hot water. Be sure the water temperature does not exceed 145 degrees Fahrenheit, it could damage the plastic parts. Finally, sanitize the ice bin, door, ice cube scoop, grid panel and grid with a bleach solution.

Reinstall all parts in the reverse order of disassembly, and test the ice machine operation. After the cleaning treatment, apply a release agent to the evaporator plate. This agent will retard any future build-up of scale and mineral deposits, and it will make the plate more slippery, which will provide for a better ice slab release.

To clean the condenser coil, remove the screws that secure the front grill and remove the grill (Fig. 14-34). Vacuum all lint and dust from the coil and from the surrounding area (Fig. 14-35). Reinstall the grill. The frequency of cleaning will be determined by the surrounding conditions.

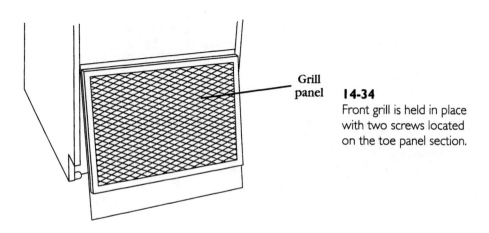

Grill panel

14-34
Front grill is held in place with two screws located on the toe panel section.

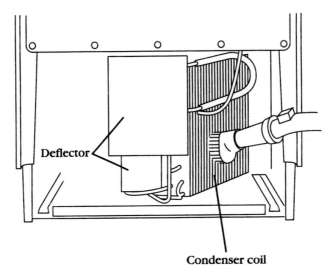

Deflector

Condenser coil

14-35 The condenser section of the ice maker.

REPAIR PROCEDURES

Each repair procedure is a complete inspection and repair process for a single ice cube maker component, containing the information you need to test and replace components.

Compressor, relay, overload protector

The typical complaints associated with compressor, relay, and overload protector failure are: 1. Ice maker does not run at all. 2. No new ice production. 3. Ice cubes in the storage bin are melting rapidly. 4. Compressor won't run, it only hums.

1. *Verify the complaint* Verify the complaint by asking the customer to describe what the ice cube maker is doing.

2. *Check for external factors* You must check for external factors not associated with the appliance. Is the appliance installed properly? Does the ice maker have the correct voltage?

3. *Disconnect the electricity* Before working on the ice maker, disconnect the electricity. This can be done by pulling the plug from the receptacle. Or disconnect the electricity at the fuse panel or at the circuit breaker panel. Turn off the electricity.

4. *Gain access to the compressor* Access the compressor. To access the compressor, remove the front grill. Remove the two screws in the condensing unit base and pull the unit toward you. Be careful to not damage any refrigerant lines. Next, remove the compressor terminal cover (Fig. 14-36) by removing the retaining clip that secures the cover. Remove the terminal cover.

5. *Test the compressor relay* To test the compressor relay, remove the relay by pulling it from the compressor terminals, without twisting it (Fig. 14-37). On the relay body is stamped the word TOP. Hold the relay so that TOP is in the up position. Next, place the probes of the ohmmeter on the relay terminals

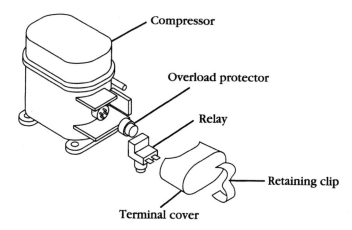

14-36 A pictorial view of compressor, overload, and relay.

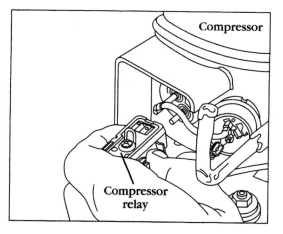

14-37
Pull the relay off the compressor. Be careful, do not break the compressor pins.

marked S and M. Set the meter scale on R × 1. The reading will show no continuity. Next, remove the probe from the terminal marked M, and place the probe on the side terminal marked L. The reading will show no continuity. Now, move the probe from terminal S, and place it on the terminal marked M. The reading will show continuity. With the probes still attached, turn the relay upside down (Fig. 13-34B) and perform the same tests. By turning the relay over, the switch contacts in the relay will close. When you retest the relay, you should get the opposite results. You should have continuity between terminals S and M, and between S and L. The meter will not show continuity between M and L. If the relay fails this test, replace it.

6. *Test the overload protector* To test the overload protector, remove the wires from the overload protector and compressor terminals. Then, remove the overload protector (from the compressor) by removing the retaining clip that secures the overload (Fig. 14-38). Next, place the probes of the ohmmeter on the overload terminals. Set the meter scale on R × 1. The reading will show continuity. If not, replace the overload protector.

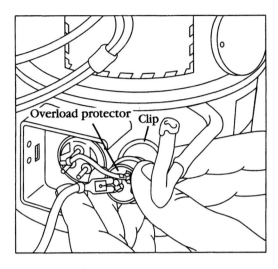

14-38
Disconnect the overload protector.

7. *Test the compressor* To test the compressor, remove the relay and the overload protector. This will expose the compressor terminals. The compressor terminals are marked C, S, and R. The C indicates the common winding terminal; the S indicates the start winding terminal; and the R indicates the run winding terminal. Next, place the probes of the ohmmeter on the terminals marked S and R (Fig. 14-39). Set the meter scale on R × 1, and adjust the needle setting to indicate a 0 reading. The meter reading will show continuity. Now, place the meter probes on the terminals marked C and S. The meter reading will show continuity. Finally, place the meter probes on the terminals marked C and R. The meter reading will show continuity. The total number of ohms measured between S and R is equal to the sum of C to S plus C to R. To test the compressor for ground, place one probe on a compressor terminal, and the other probe on the compressor housing, or on any good ground (Fig. 14-40). Set the meter scale to R × 1000.

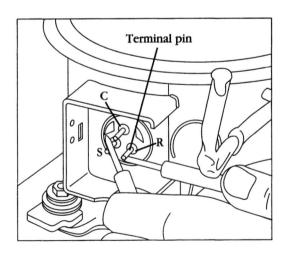

14-39
Testing the compressor windings.

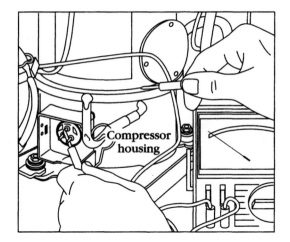

14-40
Testing the compressor for ground.

The meter reading will show no continuity. Repeat this for the remaining two terminals. The meter reading will show no continuity. If you get a continuity reading from any of these terminals to ground, the compressor is grounded, and it must be replaced.

Bin thermostat

The typical complaints associated with bin thermostat failure are: 1. The ice maker runs all the time, making too much ice. 2. The ice maker doesn't run at all.

1. *Verify the complaint* Verify the complaint by checking the sensing tube and bulb and the control settings.
2. *Check for external factors* You must check for external factors not associated with the appliance. Is the appliance installed properly? Explain to the user how to set the controls.
3. *Disconnect the electricity* Before working on the ice maker, disconnect the electricity. This can be done by pulling the plug from the receptacle. Or disconnect the electricity is at the fuse panel or at the circuit breaker panel. Turn off the electricity.
4. *Gain access to the bin thermostat* To access the bin thermostat, remove the screws from the escutcheon (Fig. 14-41) and remove the panel. Next, remove the screws from the control bracket (Fig. 14-42). Pull back on the control bracket, exposing the controls.
5. *Test the bin thermostat* To test the bin thermostat, remove the wires from the thermostat terminals, and place the ohmmeter probes on those terminals (Fig. 14-43). Set the range scale on R × 1, and test for continuity. The meter should read continuity between the contacts, if the temperature of the capillary tube is above 42 degrees Fahrenheit. The meter should not read continuity between the contacts, if the temperature of the capillary tube is below 36 degrees Fahrenheit.
6. *Remove the bin thermostat* Remove the bin thermostat from the control bracket by removing the two screws (Fig. 14-41). Remove well from the left wall of the liner. Next, remove the five clips (under the gasket) from the

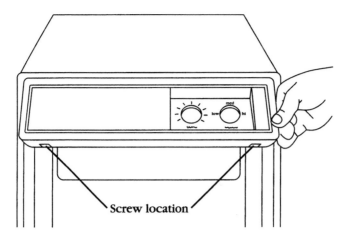

14-41 Removing the control panel to expose the controls.

left-hand side of the liner (Fig. 14-44). Now, bend the liner flange forward and remove the capillary tube and thermostat control.

7. *Install a new bin thermostat* To install the new bin thermostat just reverse the order of disassembly, and reassemble. Then, test the control. Remember to reinstall the capillary tube in the same location from where it was removed. If you do not, the ice maker will not cycle properly.

Evaporator thermostat

The typical complaints associated with evaporator thermostat failure are: 1. Unable to control ice cube thickness. 2. Ice maker water not circulating. 3. The ice maker runs, but no ice in the bin.

1. *Verify the complaint* Verify the complaint by checking the sensing tube and bulb, and the control settings.

2. *Check for external factors* You must check for external factors not associated with the appliance. Is the appliance installed properly? Explain to the user how to set the controls.

3. *Disconnect the electricity* Before working on the ice maker, disconnect the electricity. This can be done by pulling the plug from the receptacle. Or disconnect the electricity at the fuse panel or at the circuit breaker panel. Turn off the electricity.

4. *Gain access to the evaporator thermostat* To access the evaporator thermostat, remove the screws from the escutcheon (Fig. 14-41), and remove the panel. Next, remove the screws from the control bracket (Fig. 14-42). Pull back on the control bracket, exposing the controls.

5. *Test the evaporator thermostat* To test the evaporator thermostat, remove the wires from the thermostat terminals, and place the ohmmeter probes on terminals 1 and 2 (Fig. 14-45). Set the range scale on R × 1, and test for continuity. The meter should show continuity between the contacts, if the

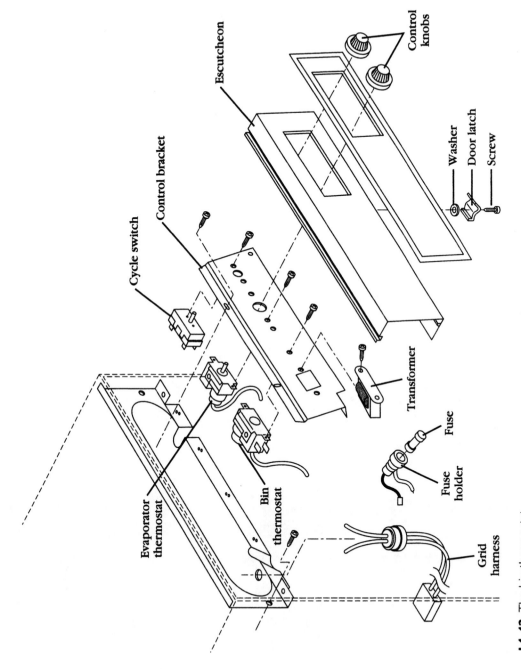

Escutcheon

Control knobs

Washer

Door latch

Screw

Control bracket

Cycle switch

Transformer

Fuse

Fuse holder

Evaporator thermostat

Bin thermostat

Grid harness

14-42 The bin thermostat.

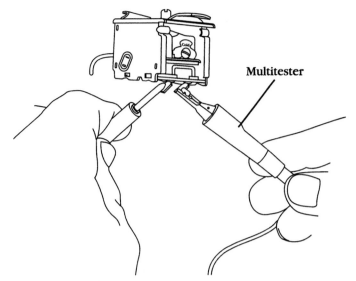

14-43 The bin thermostat.

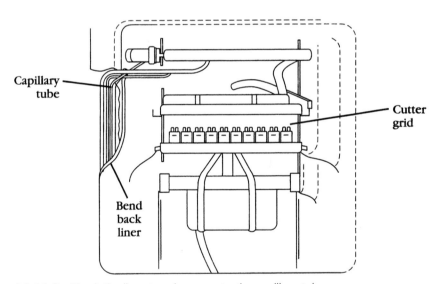

14-44 Peel back the liner to gain access to the capillary tube.

temperature of the evaporator freeze-plate is 30 degrees Fahrenheit or warmer. The meter should not read continuity between the contacts, if the temperature of the evaporator freeze-plate is +10 to –3 degrees Fahrenheit. By disconnecting the water pump at the terminal board, and operating the ice maker without the pump, the evaporator thermostat action might be easily observed. This will cause the thermostat to cycle in a matter of a few minutes.

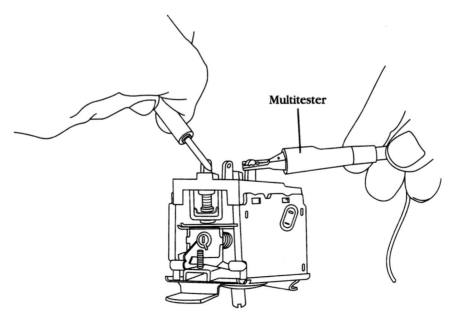

Multitester

14-45 The evaporator thermostat.

6. *Remove the evaporator thermostat* Remove the cutter grid by removing the thumbscrews (Fig. 14-32). Remove the evaporator thermostat from the control bracket by removing the two screws (Fig. 14-42). Remove the clamp from underneath the evaporator freeze-plate, which secures the capillary tube to the evaporator. Next, remove the five clips, under the gasket, from the left-hand side of liner (Fig. 14-44). Now, bend the liner flange forward and remove the capillary tube and thermostat control.

7. *Install a new evaporator thermostat* To install the new evaporator thermostat, just reverse the order of disassembly, and reassemble. Then, test the control. Remember to reinstall the capillary tube in the same location from where it was removed. Also, the capillary tube must be taped to the hot gas restrictor tube. If you do not, the ice maker will not cycle properly.

Hot gas solenoid valve

The typical complaints associated with hot gas solenoid valve failure are: 1. Ice maker runs, but there is no ice production. 2. Evaporator freeze-plate will not heat up to release ice slab. 3. Ice maker runs continuously.

1. *Verify the complaint* Verify the complaint by checking the ice maker cycles.

2. *Check for external factors* You must check for external factors not associated with the appliance. Is the appliance installed properly? Explain to the user how to set the controls.

3. *Disconnect the electricity* Before working on the ice maker, disconnect the electricity. This can be done by pulling the plug from the receptacle. Or disconnect the electricity at the fuse panel or at the circuit breaker panel. Turn off the electricity.

4. *Gain access to the hot gas solenoid valve* To access the hot gas solenoid valve (Fig. 14-46), remove the grill (Fig. 14-34). Next, remove the deflector from the condenser (Fig. 14-35).

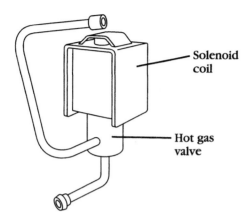

Solenoid coil

Hot gas valve

14-46
The hot gas defrost valve. This valve will reverse the flow of refrigerant to the evaporator in the defrost cycle.

5. *Test the hot gas solenoid valve* Test the hot gas solenoid valve for continuity. Remove the wires from the solenoid coil. Place the ohmmeter probes on the solenoid coil terminals (Fig. 14-46). Set the range scale on R × 1, and test for continuity. To test the hot gas valve itself, connect a 120-volt fused service cord (Fig. 14-47) to the solenoid coil. Listen for a click sound as the plunger raises up. Now, disconnect the service cord, and you will hear the plunger drop back. If you cannot hear a distinctive click sound from the hot gas valve, then it will need to be replaced by an authorized service company (the sealed system might be under warranty from the manufacturer) or by a licensed refrigerant technician.

The solenoid coil is a separate component that can be replaced without replacing the entire hot gas valve assembly. Another way to test the hot gas valve is to leave the wires off the solenoid coil, and reconnect the service cord to the solenoid coil. This test requires the electricity to be turned on.

Caution: Tape the solenoid coil leads that were removed so that they will not touch the chassis when you plug in the ice maker for this test. Be cautious when working with live wires. Avoid getting shocked!

With the ice maker plugged in and running, feel the hot gas defrost tube, it should feel warm, or hot, when the valve is energized.

6. *Remove the hot gas solenoid coil* To remove the hot gas solenoid coil, remove the spring clip from the top of the coil, and remove the coil (be sure the electricity is off).

7. *Install a new hot gas solenoid coil* To install the new solenoid coil, just reverse the order of disassembly, and reassemble. Then, test the valve.

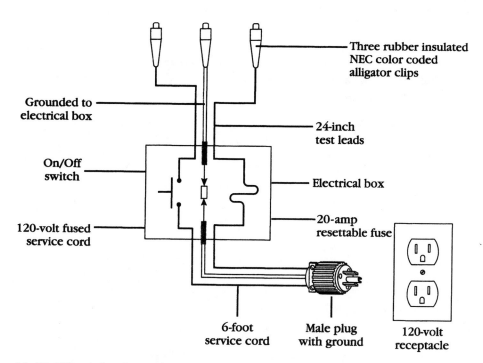

14-47 120-volt fused service test cord.

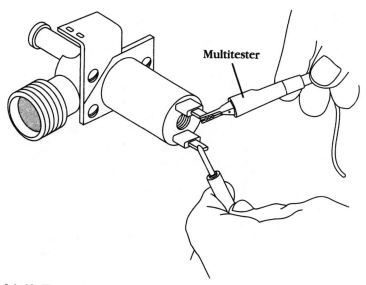

14-48 The water valve.

Water valve

The typical complaints associated with water valve failure are: 1. Ice maker runs, but there is no ice production. 2. No water circulating across the evaporator freeze-plate. 3. Water flooding the storage bin, causing the ice to melt.

1. *Verify the complaint* Verify the complaint by checking the ice maker cycles.
2. *Check for external factors* You must check for external factors not associated with the appliance. Is the appliance installed properly? Explain to the user how to set the controls. Is the water supply turned on?
3. *Disconnect the electricity* Before working on the ice maker, disconnect the electricity. This can be done by pulling the plug from the receptacle. Or disconnect the electricity at the fuse panel or at the circuit breaker panel. Turn off the electricity.
4. *Gain access to the water valve* To access the water valve, remove the top insulated panel. The water valve is located at the top right front corner.
5. *Remove and test the water valve* In order to test the water valve solenoid coil, the water valve must be removed from the storage bin. Shut off the water supply to the ice maker. Now, disconnect the water line from the valve. Next, remove the screws from the water valve bracket. Pull on the valve to release it from the receptacle in the liner. Place the ohmmeter probes on the solenoid coil terminals (Fig. 14-48). Set the range scale on R × 1, and test for continuity. If there is no continuity, replace the water valve.
6. *Install a new water valve* To install the new water valve, just reverse the order of disassembly, and reassemble. Then, test the valve. Don't forget to turn on the water supply.

Condenser fan motor

The typical complaints associated with condenser fan motor failure are: 1. The ice maker stopped producing ice. 2. The condenser fan motor runs slower than normal. 3. The condenser fan motor does not run at all. 4. The compressor is sometimes noisier than normal.

1. *Verify the complaint* Verify the complaint by asking the customer to describe what the ice maker is doing. Is the condenser fan motor running during the freeze cycle?
2. *Check for external factors* You must check for external factors not associated with the appliance. Is the appliance installed properly? Are there any foreign objects blocking the condenser fan blades?
3. *Disconnect the electricity* Before working on the ice maker, disconnect the electricity to the ice maker. This can be done by pulling the plug from the receptacle. Or disconnect the electricity at the fuse panel or at the circuit breaker panel. Turn off the electricity.
4. *Gain access to the condenser fan motor* To access the condenser fan motor, remove the front grill. Remove the two screws in the condensing unit base and pull the unit toward you. Be careful not to damage any refrigerant lines.

5. *Test the condenser fan motor* To test the condenser fan motor, remove the wires from the motor terminals. Next, place the probes of the ohmmeter on the motor terminals (Fig. 14-49). Set the meter scale on R × 1. The meter should show some resistance. If no reading is indicated, replace the motor. If the fan blades do not spin freely, replace the motor. If the bearings are worn, replace the motor.

6. *Remove the condenser fan motor* To remove the condenser fan motor, you must first remove the fan blades. Unscrew the nut that secures the blades to the motor. Remove the blades from the motor. Then, remove the motor assembly by removing the mounting bracket screws (Fig. 14-50).

7. *Install a new condenser fan motor* To install the new condenser fan motor, just reverse the order of disassembly, and reassemble. Remember to reconnect the ground wire to the motor. Reconnect the wires to the motor terminals, and test.

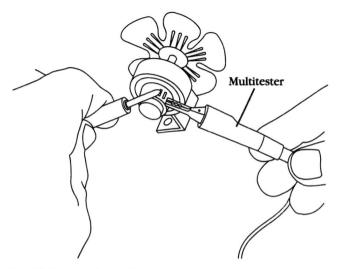

14-49 The condenser fan motor.

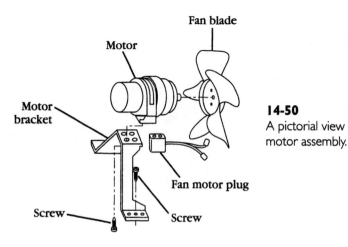

14-50
A pictorial view of the condenser fan motor assembly.

Water pump

The typical complaints associated with water pump failure are: 1. Ice maker runs, but there is no ice production. 2. No water circulating across the evaporator freeze-plate.

1. *Verify the complaint* Verify the complaint by checking the ice maker cycles.
2. *Check for external factors* You must check for external factors not associated with the appliance. Is the appliance installed properly? Explain to the user how to set the controls.
3. *Disconnect the electricity* Before working on the ice maker, disconnect the electricity. This can be done by pulling the plug from the receptacle. Or disconnect the electricity at the fuse panel or at the circuit breaker panel. Turn off the electricity.
4. *Gain access to the water pump* To access the water pump, remove the bin door, the front insulated panel, and the inner bin door. A drain plug is located under the water tank. Remove it to drain the water out of the tank (Fig. 14-33). Next, remove the thumbscrews that secure the water tank and remove the tank.
5. *Test the water pump* To test the water pump motor, isolate the motor, and place the probes of the ohmmeter on the motor terminals (Fig. 14-51). Set the meter scale on R × 1. The meter should show some resistance. If no reading is indicated, replace the water pump. Next, check the motor with a 120-volt fused service cord (Fig. 14-47).
6. *Remove the water pump* To remove the water pump (Fig. 14-52), remove the screws from the water pump bracket that secure the pump to the liner. Disconnect the discharge hose from the pump. Remove the water pump.
7. *Install a new water pump* To install new water pump, just reverse the order of disassembly, and reassemble. Reconnect the wires to the motor terminals, and test.

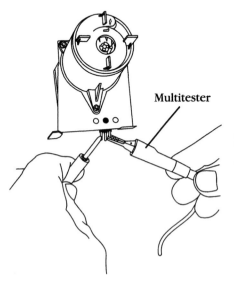

Multitester

14-51
A bottom view of water pump. When checking the pump, be sure the inlet is free of debris.

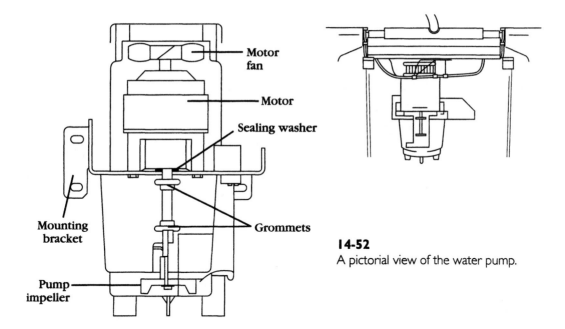

Motor
fan

Motor

Sealing washer

Mounting
bracket

Grommets

14-52
A pictorial view of the water pump.

Pump
impeller

Cutter grid

The typical complaints associated with cutter grid failure are: 1. Ice slabs lay on top of the cutter grid. 2. Cutter grid is not cutting ice slab into cubes evenly.

1. *Verify the complaint* Verify the complaint by checking the ice slab and cutter grid fuse.
2. *Check for external factors* You must check for external factors not associated with the appliance. Is the appliance installed properly? Explain to the user how to set the controls.
3. *Disconnect the electricity* Before working on the ice maker, disconnect the electricity. This can be done by pulling the plug from the receptacle. Or disconnect the electricity at the fuse panel or at the circuit breaker panel. Turn off the electricity.
4. *Gain access to the cutter grid* To access the cutter grid, open the bin door, and remove the cutter grid by removing the two thumb screws. Unplug the cutter grid, and remove it from the storage bin (Fig. 14-32).
5. *Test the cutter grid* Examine the cutter grid for broken wires, and check the connecting pins for corrosion (Fig. 14-53). As you inspect the cutter grid, look for cracked or broken insulators in the frame. Next, place the probes of the ohmmeter on the cutter grid plug terminals (Fig. 14-54). Set the meter scale on R × 1. The meter should show continuity. If no reading is indicated, one or more grid wires or insulators are defective.
6. *Repair the cutter grid* If the cutter grid frame and insulators are broken, then it would be advisable to replace the entire cutter grid. Using a C-clamp, compress the spring clip to relieve the tension (Fig. 14-55). Next, use a pair of pliers to compress the adjacent spring clip and remove the buss bar. Do

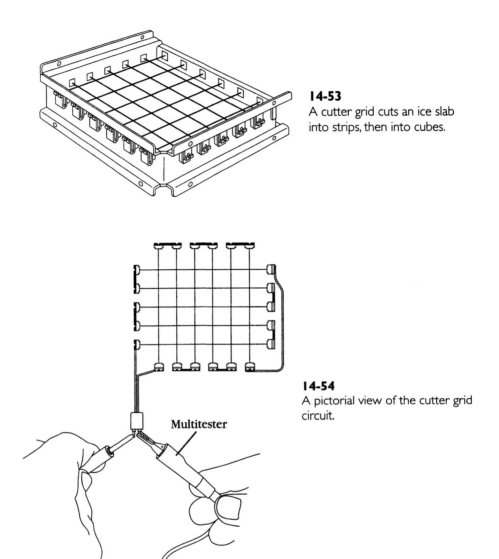

14-53
A cutter grid cuts an ice slab into strips, then into cubes.

14-54
A pictorial view of the cutter grid circuit.

Multitester

the same procedure for the other side of the cutter grid. The insulators, clips, and grid wires can now be removed and replaced. If any of the grid wires break, then, it is time to replace all of the grid wires.

7. *Install a cutter grid* To install a cutter grid, just reverse the order of disassembly, and reassemble. Reconnect the wires to the cutter grid and test it.

Cutter grid transformer and fuse

The typical complaints associated with grid transformer or fuse failure are: 1. Ice slabs lay on top of the cutter grid. 2. Cutter grid not cutting ice slab into cubes evenly.

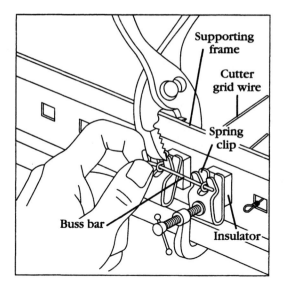

14-55
A side view of the cutter grid, showing the clips, buss bars, and insulators.

1. *Verify the complaint* Verify the complaint by checking the ice slab and cutter grid.
2. *Check for external factors* You must check for external factors not associated with the appliance. Is the appliance installed properly?
3. *Disconnect the electricity* Before working on the ice maker, disconnect the electricity. This can be done by pulling the plug from the receptacle. Or disconnect the electricity at the fuse panel or at the circuit breaker panel. Turn off the electricity.
4. *Gain access to the cutter grid transformer and fuse* To access the cutter grid transformer and fuse, remove the screws from the escutcheon (Fig. 14-41), and remove the panel. The fuse is located on the left front. To remove the fuse, push in and twist the fuse, it will pop out of the holder (Fig. 14-56). Next, remove the screws from the control bracket (Fig. 14-42). Pull back on the control bracket, exposing the controls.
5. *Test the cutter grid transformer and fuse* To test the transformer, disconnect the wires from the transformer, to isolate it from the circuit. Use a 120-volt fused service cord, and connect it to the primary side of the transformer. This test requires the electricity to be turned on.

   Be cautious when working with live wires. Avoid getting shocked!

   You might have to look at the wiring diagram for assistance to identify the primary side, and for the proper color coding of the wires. Using the volt meter, connect the probes to the secondary side of the transformer. Plug in the 120-volt fused service cord. The meter should read 8.5 volts. Next, unplug the service cord.

   To test for resistance, disconnect the power cord, and set the ohmmeter scale on R × 1. Place the probes on the primary wires of the transformer. The meter should show resistance. Next, place the probes on the secondary wires of the transformer (Fig. 14-57). The meter should show resistance. If the transformer fails either test, replace it.

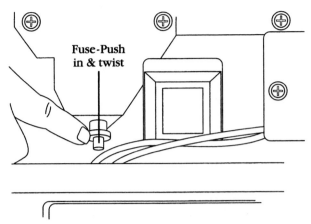

14-56 The cutter fuse is located behind control panel.

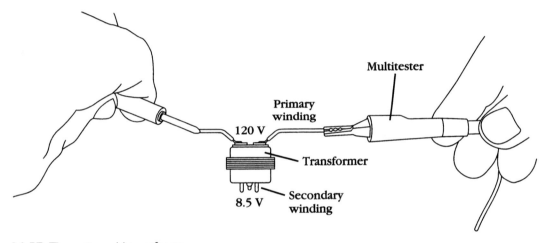

14-57 The cutter grid transformer.

To test the fuse, place the probes on each end of the fuse (Fig. 14-58). Set the ohmmeter scale on R × 1. The meter should show continuity. If not, replace the fuse.

6. *Remove the transformer* To remove the transformer, remove the screws that secure the transformer to the control bracket (Fig. 14-42).

7. *Install a new transformer or fuse* To install the transformer or fuse, just reverse the order of disassembly, and reassemble. Reconnect the wires to the components and test it.

Cycle switch

The typical complaints associated with cycle switch failure are: 1. Unable to turn on the clean cycle. 2. Unable to turn on the ice maker.

1. *Verify the complaint* Verify the complaint by checking the control settings.

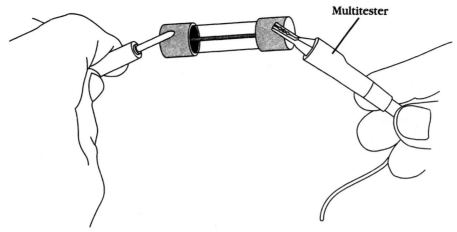

Multitester

14-58 The cutter grid fuse. If this fuse blows, the cutter grid will not function. Check for a short.

2. *Check for external factors* You must check for external factors not associated with the appliance. Is the appliance installed properly? Explain to the user how to set the controls.

3. *Disconnect the electricity* Before working on the ice maker, disconnect the electricity. This can be done by pulling the plug from the receptacle. Another way to disconnect the electricity is at the fuse panel, or at the circuit breaker panel. Turn off the electricity.

4. *Gain access to the cycle switch* To access the cycle switch, remove the screws from the escutcheon (Fig. 14-41), and remove the panel. Next, remove the screws from the control bracket (Fig. 14-42). Pull back on the control bracket, exposing the controls.

5. *Test the cycle switch* To test the cycle switch for continuity between the switch contacts, refer to the wiring diagram for the correct switch positions (Fig. 14-59). Disconnect the wires from the switch, and turn the switch to the "on" position. Place the probes of the ohmmeter on the cycle switch terminals marked 2 and 3. Set the meter scale on R × 1. The meter should show continuity. Now, place the probes of the ohmmeter on the cycle switch terminals marked 1 and 6. The meter should show continuity. Next, turn the cycle switch to the clean position. Place the probes of the ohmmeter on the cycle switch terminals marked 5 and 6. The meter should show continuity. If the switch fails these tests, replace it.

6. *Remove the cycle switch* To remove the cycle switch, remove the two screws that secure the switch to the control bracket (Fig. 14-42).

7. *Install a new cycle switch* To install a new cycle switch, just reverse the order of disassembly, and reassemble. Connect the wires to the switch terminals according to the wiring diagram and test it.

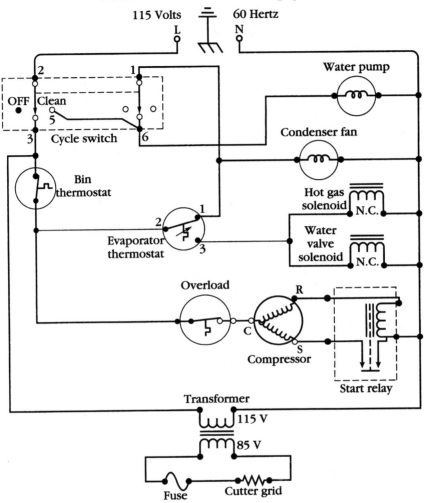

14-59 A sample wiring diagram of a self-contained ice cube maker.

Glossary

allen wrench An "L"-shaped tool that is used to remove hex screws.

alternating current (ac) Electrical current that flows in one direction, then reverses itself and flows in the opposite direction. In 60-cycle current, the direction of flow reverses every 120th second.

ambient temperature Temperature of air that surrounds an object on all sides.

ammeter A test instrument used to measure current.

ampere The number of electrons passing a given point in one second.

armature The section of a motor that turns.

automatic defrost timer A device, connected in an electrical circuit, which shuts off the refrigeration cycle, and turns on the electric heaters to melt the ice off the evaporator coil.

bake element Lower heating element in an oven, used for baking foods.

belt A band of flexible material used to transfer mechanical power from one pulley to another pulley.

bimetal (strip or thermostat) Two dissimilar metals joined together to form one unit with a differential expansion rating, and it will bend if there is a temperature change.

bracket A hardened structure used to support a component.

broil element Upper heating element in an oven used for broiling foods.

cabinet The outer wrapper of an appliance.

cam A rotating surface with rises and falls that opens and closes the switches in a timer.

capacitor A device that stores electricity, used to start and/or run circuits on many large electric motors.

capillary tube A metering device used to control the flow of refrigerant in a sealed system. It usually consists of several feet of tubing with a small inside diameter.

Celsius The metric system for measuring temperature. The intervals between the freezing point and the boiling point of water is divided into 100 degrees. This is also called the *centigrade scale*.

circuit A path for electrical current to flow from the power source, to the point of use, and back to the power source.

circuit breaker A safety device used to open a circuit if that circuit is overloaded.

closed (circuit) An electrical circuit in which electrons are flowing.

component An individual mechanical or electrical part of an appliance.

compressor An apparatus that is similar to a pump used to increase the pressure of refrigerant that is to be circulated within a closed system.

condenser coil A component of the refrigeration system that transfers the heat that came from within a closed compartment to the outside surrounding area.

condenser fan A fan motor and fan blades used to cool the condenser coil.

contact points Two movable objects, or contacts, that come together to complete a circuit; or to separate and break a circuit. These contact points are usually made of silver.

continuity The ability of a completed circuit to conduct electricity.

control A device, either automatic or manual, that is used to start, stop, or regulate the flow of liquid, gas, or electricity.

current The flow of electrons from a negative to a positive potential.

cycle A series of events that repeat themselves in the same order.

defective A component that does not function properly.

defrost cycle That part of the refrigeration cycle in which the ice is melted off the evaporator coil.

defrost timer A device used in an electrical circuit to turn off the refrigeration long enough to permit the ice to be melted off the evaporator coil.

diagnosis The act of identifying the problem from its signs and symptoms.

direct current Electric current that flows in one direction in a circuit.

end bell The end plates that hold a motor together.

energize To supply electrical current for operation of an electrical component.

evaporator coil A component of the refrigeration system that removes the heat from within a closed compartment.

evaporator fan A motor and fan blades used to circulate the cold air.

Fahrenheit The standard system for measuring temperature. The freezing point of water is 32 degrees and the boiling point of water is 212 degrees.

fuse A safety device used to open a circuit if that circuit is overloaded.

gasket A flexible material used to seal components together; either air-tight or water-tight.

ground A connection to earth or to another conducting body that transmits current to earth.

ground wire An electrical wire that will safely conduct electricity from a structure to earth.

Hertz (Hz) A unit of measurement for frequency. One hertz equals one cycle per second.

hot gas defrost A defrosting system in which hot refrigerant from the condenser coil is directed to the evaporator coil, for a short period of time, to melt the ice from the evaporator coil.

housing A metal or plastic casing that covers a component.

idler pulley A device that rests on or presses against a drive belt to maintain a specified tension on the drive belt.

insulation Substance used to retard or slow the flow of heat through a substance.

insulator A material that does not conduct electricity. It is used to isolate current-carrying wires or components from other metal parts.

ladder diagram A wiring schematic, where all of the components are stacked in the form of a ladder.

lead (wire) A section of electrical wiring that is attached to a component.

lint Fine pieces of cotton fiber that broke away from the garment.

module A device that is self contained with a group of interconnecting parts designed to do a specific job.

nut driver A tool used to remove or reinstall hexagonal-head screws or nuts.

Ohm A unit of measurement for electrical resistance.

ohmmeter A test instrument used for measuring resistance.

open (circuit) A break in an electrical circuit that stops the flow of current.

overload protector A device that is either temperature, pressure, or current operated, and is used to open a circuit to stop the operation of that circuit if dangerous conditions should arise.

parallel circuit Components that are parallel-connected across one voltage source. All of the branches are supplied with the same amount of voltage.

pressure switch A device that is operated by pressure, and turns a component on or off.

pulley A wheel turned by, or driving, a belt.

refrigerant A chemical substance, used in refrigeration and air conditioning, that produces a cooling effect.

relay A magnetic switch which uses a small amount of current, in the control circuit, to operate a component needing a larger amount of current in the operating circuit. A remote switch.

relief valve A safety device that is designed to open before dangerous pressure is reached.

resistance The opposition to current flow. The load in an electrical circuit.

run winding Electrical winding of a motor, which has current flowing through it during normal operation of motor.

safety thermostat A thermostat that limits the temperature to a safe level.

schematic diagram A line drawing which gives the electrical paths, layout of components, terminal identification, color codes of wiring (and sometimes the sequence of events) of an appliance.

series circuit A circuit where all of the components are sequentially connected in the same line. If one component fails, all the components fail.

short circuit An electrical condition where part of a circuit errantly touches another part of a circuit, and causes part or all of the circuit to fail (and trip the circuit breaker or blow a fuse).

solenoid A cylindrical coil of insulated wire that establishes a magnetic field in the presence of current.

start winding A winding in an electric motor used only during brief periods when the motor is starting.

surface element Top cooking element on a range used for cooking foods.

switch A device to turn current on or off in an electrical circuit.

temperature A measure of heat energy, or the relative lack thereof.

terminal A connecting point in a circuit to which a wire would be attached to connect a component.

test light A light provided with test leads that is used to test electrical circuits.

thermometer A device used to measure temperature.

thermostat A device that senses temperature changes, and which usually operates a control relay.

thermostat, operating A thermostat that controls the operating temperature of a component.

transformer A device that raises or lowers the main ac supply voltage.

voltage The difference in potential between two points; or the difference in static charges between two points.

voltmeter A test instrument used to measure voltage.

VOM A test instrument used to measure voltage, resistance, and amperage. A volt-ohm-milliammeter.

watt A unit to measure electrical power.

wattmeter A test instrument used to measure electrical power.

Index

Illustration page numbers are in **boldface**.

I

ice makers (*see also* refrigerators & freezers), 357-404
  diagnostic chart, 368, **368**
  ejector blade position, 362, **363**
  empty cube compartments, 367-368
  flooding, 366
  frozen fill-tube, 374
  half-filled cube tray, 374
  head and tray, 375-377, **378**
  heater assembly, 373-374, **374**
  hollow cubes, 366, 370
  inoperative: no ice, 365-366, 367, **368**, 369, 370, 373, 374, 375, 377
  low ice production, 365-366
  mechanical linkage, 99, **100**
  moldy cubes, 373
  no water, 367
  operation of ice makers, 358-360
  removal procedure, 369, **369**, **370**, **371**
  safe operation/service, 357
  self-contained units, 379-404, 379
    clean-cycle inoperative, 402
    cleaning, 384-385
    compressor, 386-389, **387**, **388**, **389**
    compressor hums: won't run, 386
    compressor inoperative, no ice, 381-382
    compressor runs continuously, ice in bin, 382
    compressor runs, no ice, 382, 390, 393, 396, 398
    condenser fan motor, 396-397, **397**
    condenser fan won't run, 383
    condenser, **386**
    continuous operation, 389, 393
    cutter grid transformer and fuse, 400-402, **401**, **402**, **403**
    cutter grid, 399-400, **400**
    cycle switch, 402-403
    disassembly, **384**
    dripping water, 382
    evaporator thermostat, 390, 392-393, **393**
    fan inoperative, 396
    fan runs slowly, 396
    flooding, 396
    front grille removal, **385**
    hot-gas solenoid valve, 393-394, **394**, **395**
    ice slab won't fall, 393, 399, 400
    inoperative: no ice, 386, 389, 396, 402
    low ice production, 382
    melting ice, 386
    noisy operation, 396
    operation, 379-381
    overload protector, 386-389, **388**
    refrigeration system, 380, **381**
    relays, 386-389, **387**
    thermostat, bin thermostat, 389-390, **391**, **392**
    thick cubes, 383, 390
    thin cubes, 383, 390
    too much ice, 389
    uneven sized cubes, 399, 400
    water not circulating, 390, 396, 398
    water pump, 398, **398**, **399**
    water pump won't run, 383
    water quality, 383-384
    water system, **380**
    water tank, **385**
    water tank empty, 383
    water valve, **395**, 396
    wiring diagram, **404**
  small cubes, 366, 367-368
  solid block of ice: no cubes, 377
  spills water from tray, 367
  stalls in cycle, 369
  stuck cubes, 373, 375, 377
  stuck-together cubes, 367
  temperature, 365
  test points, shorting out, 362, **363**
  testing type 1 ice makers, 360-362, **361**, **362**
  testing type 2 ice makers, 363-364, **364**, **365**
  thermostat, 370-372, **371**, **372**
  too much ice, 365
  tray failure, 377-379, **378**
  twist-mode for ice cube tray, **360**
  type 1 vs. type 2 ice makers, 358-360, **359**, **360**, **361**, **362**
  water quality problems, 364-365
  water supply, **358**
  water valves, 93-94, **93**, **94**, **95**, 374-375, **375**, **376**, **377**
installation tips
  dishwashers, 106-108, **109**, **110**
  dryers, 227-228, **228**
  freestanding range, 3-4
  garbage disposers, 155, **156**
  ranges, 265
  washers, 187

L

ladder diagrams, 69-70, **70**
  dryers, **257**
  washers, **224**
loads, electrical, 50

M

Major Appliance Consumer Action Panel (MACAP), 24-27, 29
manuals, 26, 28-29

About the author

Eric Kleinert is a professional with 24 years experience in commercial and domestic major appliance, refrigeration, and HVAC sales, service, and installation. He owns and operates a major appliance and air conditioning sales and service corporation. Also, as an instructor on behalf of the Palm Beach County School District in Florida, he teaches adults aspects of preventative and diagnostic services and techniques, which enhance their ability to better evaluate products and services necessary to maintain residential climate-control systems and major appliances.